DIMINISHING RETURNS
AT WORK

Diminishing Returns at Work

The Consequences of Long Working Hours

John H. Pencavel

Oxford University Press is a department of the University of Oxford. It furthers
the University's objective of excellence in research, scholarship, and education
by publishing worldwide. Oxford is a registered trade mark of Oxford University
Press in the UK and certain other countries.

Published in the United States of America by Oxford University Press
198 Madison Avenue, New York, NY 10016, United States of America.

© Oxford University Press 2018

All rights reserved. No part of this publication may be reproduced, stored in
a retrieval system, or transmitted, in any form or by any means, without the
prior permission in writing of Oxford University Press, or as expressly permitted
by law, by license, or under terms agreed with the appropriate reproduction
rights organization. Inquiries concerning reproduction outside the scope of the
above should be sent to the Rights Department, Oxford University Press, at the
address above.

You must not circulate this work in any other form
and you must impose this same condition on any acquirer.

Library of Congress Cataloging-in-Publication Data
Names: Pencavel, John H., author.
Title: Diminishing returns at work : the consequences of long
working hours / John H. Pencavel.
Description: New York : Oxford University Press, 2018.
Identifiers: LCCN 2017052160 | ISBN 9780190876166 (hardback) |
ISBN 9780190876173 (UPDF) | ISBN 9780190876180 (epub)
Subjects: LCSH: Hours of labor—United States. | Hours of labor—Great
Britain. | Wages—United States. | Wages—Great Britain. | Labor
productivity—United States. | Labor productivity—Great Britain. |
Industrial hygiene—United States. | Industrial hygiene—Great Britain. |
BISAC: BUSINESS & ECONOMICS / Labor. | BUSINESS & ECONOMICS /
Workplace Culture. | BUSINESS & ECONOMICS / Corporate & Business History.
Classification: LCC HD5124 .P46 2018 | DDC 331.25/70941—dc23
LC record available at https://lccn.loc.gov/2017052160

This book is dedicated to my American and English families.

Contents

Preface and Acknowledgments | ix

1. Introduction: Why Working Hours? | 1

2. A Brief History of Working Hours | 13
   Trade Unions | 17
   Employers | 25
   Statutory Legislation | 30
   A Market Explanation | 38
   America and Britain Contrasted | 42
   What Has Been Learned? | 48

3. Conceptual Framework | 52
   An Expression Inspired by Charles Cobb and Paul Douglas | 57
   An Expression Inspired by Benjamin Gompertz | 62

4. Estimates of Production Functions | 68
   Observations Collected by Horace Vernon During the First World War | 68
   Observations Collected by the Industrial Health Research Board During the Second World War | 95

Observations Collected by Max Kossoris and Reinfried Kohler During the Second World War | 104
Observations Collected by Ben Craig on Plywood Mills in the State of Washington | 111
Some Conclusions About Hours and Output of Workers from These Four Cases | 125

5. Further Implications of the Augmented Production Functions | 133
Part-Time and Full-Time Workers | 133
Overtime Work | 136
Macroeconomics | 137

6. Hours of Work, Health, and Well-Being | 138
Cognitive Function | 143
Cardiovascular Disease | 144
Nationally Representative Populations | 145
Injuries and Accidents | 148
The Well-Being of the Household | 149

7. The Association Between Working Hours and Hourly Earnings | 151
The Demise of a Basic Issue | 151
Information Provided by Self-Employed Workers | 172
A Third Class of Explanations | 180

8. Concluding Notes | 182

DATA APPENDICES | 193
REFERENCES | 227
INDEX | 243

Preface and Acknowledgments

I have conducted research on working hours for years. Over these years, I have received help from many people—students, colleagues, researchers, referees—in the form of constructive criticism, of collaboration in the research, and of data collection and statistical analysis. I thank them for their help. I thank Louise Pencavel and Norma Virgoe, who made significant contributions to this book.

This book stems from articles that I have written in recent years and that are listed in the references. Although some of the observations here have been used in previous work, none of the equations specified in chapter 3 and estimated in chapter 4 has appeared elsewhere. All were calculated by Katy Bergstrom, an outstanding research assistant. I thank her for her work.

Research support from the Russell Sage Foundation (Award #: 85-16-05) is acknowledged with thanks.

I also acknowledge those scholars who have expressed (at least implicitly) many of the arguments presented in this book. An incomplete list is Robert A. Hart (1987), Chris Nyland (1989),

John Rae (1894), Marcus Rubin and Ray Richardson (1997), John Treble (2001), and Michael White (1987).

My editor at Oxford University Press, David Pervin, shepherded the manuscript as it passed through its various stages, and he supported its preparation. Thank you.

DIMINISHING RETURNS AT WORK

1

Introduction: Why Working Hours?

A common gauge of an individual's material standard of living is his or her income or consumption over a period of time. Measuring an individual's well-being in this way neglects the means by which this income or consumption has been obtained. Yet, for most people, income is acquired by selling their time and lending their efforts to someone else. Part of this time and energy may not be entirely unpleasant, but many would prefer to spend less time in this way, even if it entails lower income.

Among those with the same income or consumption level, the individual who works longest for that income would be judged by many as having the lowest well-being. This judgment helps to explain why some societies have applied a higher tax rate to nonlabor income than to labor income. Thus, the hours that people work in labor markets are linked to their standard of living.

Working hours also matter to employers, as testified by their willingness to spend resources to resist any reductions in hours worked by their employees. The employer determines the work schedule for employees. The manner in which these working hours (together with other inputs) generate an output is embodied in the firm's production function. It is often assumed the employer has complete knowledge of the production function in the workplace, although the history of working hours might cause one to question this assumption.

The production function is one of the basic constructs of economics. It describes the manner in which factors or inputs (such as land, labor, raw materials, and physical capital) are converted by an organization into one or several outputs. This monograph is concerned with the input of labor in production and, in particular, with the input and consequences of hours of work.

The marginal product of working hours in a production function may be defined as the change in output as a result of an additional hour of work, when all other inputs are fixed. When working hours are useful in producing output, an additional hour worked (when other inputs are constant) does not cause output to fall; more likely output will rise. So, normally, the marginal product of working hours is positive.

Usually economists assume that the marginal product of any input is not a fixed number, that it varies with the level at which other inputs are used. Also, the marginal product of an input is assumed to vary as additional units of that input are applied in production. In particular, the law of diminishing returns—one of the basic postulates of economics—concerns the change in the marginal product of an input as more of that input is used. Applied to working hours per worker, the law of diminishing returns states that, provided all other inputs are constant, as hours of work increase, after a certain level of working hours the marginal product of hours falls. It is the marginal product of hours that lies behind the firm's demand for hours of work in orthodox models of the purposive firm; that is, according to the law of diminishing returns, after a certain number of hours, the marginal product of hours per worker falls as the number of hours increases and, correspondingly, the demand for hours per worker falls as the number of hours increases.

Introduction: Why Working Hours?

The empirical research presented in this book can be viewed as a determination of whether the law of diminishing returns applies to hours of work. Some may question the need to ascertain whether diminishing returns apply to working hours, although as Schumpeter (1954) remarks, there is no axiom implying the law, and we have a "duty of factual verification" (1037). Indeed, the validity of the law of diminishing returns is consequential: many theorems in conventional economic analysis require its satisfaction. It also appears in second-order conditions in many optimizing models of the firm. Stigler (1941) provides an account of the role of the law of diminishing returns in economic thought. In addition, the law of diminishing returns is relevant for assessing the consequences of mandatory reductions in working hours.

Not only have economists overlooked the need to confirm the law of diminishing returns as applied to working hours, but also there has been an unwitting tendency to assume it does not apply. For instance, much of what has gone under the name of "labor supply analysis" in the past forty years or so has unthinkingly assumed that diminishing returns to hours of work do not operate; this will be shown below. If the law of diminishing returns does apply to working hours, then the usual procedures followed by economists to account for differences in the hours of work among individuals are questionable. This topic is taken up in chapter 7.

It will soon become clear that much of the research described herein is comparative. Information defined by its time and place is often better digested when contrasted with the analogous information at another time or in another place. This is certainly the case for the empirical work I discuss in the chapters that follow. I concentrate on the two societies I know best: Britain and the United States of America (referred to in the text simply as

America), so there is an explicit geographical dimension to the research presented. In addition, there is a clear historical dimension in that I draw upon information about working hours in the past, as well as those in the present.

The subject of time worked in labor markets has several elements: hours worked per day, days worked per week, weeks worked per year, and years worked in a lifetime. This monograph does not cover all these components. Therefore this study is not a comprehensive analysis of time spent at work. Weight is given to hours and days of work per week and per year. Even limited in this way, there are some important distinctions. Typically, in those economies in which most workers are not self-employed, each employer schedules the hours and days of work, although actual hours may stray from the scheduled hours; workers may shirk or waste time, with the consequence that an employee's time at the workplace is not synonymous with time actually working. When employees are paid by their input of time (by the hour or by the week), they are inclined to see their earnings as related to the length of their time at the workplace, but that is poorly linked to how hard they work. Working harder is viewed as raising output, which benefits the employer rather than the employee. The employer's response is to hire supervisors who can ensure that workers are applying at least some minimum effort to the job. With employees hesitant to expend much energy at work and with supervisors paid to prevent and detect indolence, the workplace is apt to be an ongoing struggle over control of work effort and the returns to such effort.

The fundamental hypothesis in this monograph is that an hour of work does not always constitute the same unit of effective working time. On the contrary, one hour of work toward the end of a long workday or workweek may represent a smaller unit of effective working time than an hour of work earlier in the

Introduction: Why Working Hours?

workday or workweek. This concept is not recognized in contemporary research by economists on production functions. When explicit account is taken of hours of work in estimating production functions, economists often combine these hours with the number of workers so as to construct worker-hours: this is the product of the average hours worked by individuals and the total number of workers or, equivalently, the total number of hours aggregated over all workers.[1]

Thus, according to this procedure, a workforce of 120 individuals, with each working 40 hours a week, represents the same effective labor as 80 individuals each working 60 hours a week; in both cases, worker-hours are 4,800 hours. If this were true, the effect of an x% reduction in average hours and the same x% increase in employment would leave output unchanged. This reasoning is used by proponents of the "lump of labor" hypothesis according to which there is a fixed amount of work to be done and hours and workers can be *perfectly* substituted to perform that work. The research reported here casts doubt on this notion because typically reductions in hours when the labor force works 35 hours a week will require larger increases in employment to yield the same output than when the workforce works 60 hours.

A principal goal of the research presented in this monograph is to document the consequences of long working hours—an issue that engaged the minds of economists a century ago but has fallen out of favor in recent decades. Researchers in other disciplines have not neglected this matter, however, and some of their work is described in chapter 6.

To what extent are long working hours a feature of the contemporary economy? In terms of the delivery of effective labor

---

[1]. Griliches (1967) provides a classic example of the use of worker-hours in estimating production functions.

TABLE 1.1 Percent Distribution of Usual Weekly Hours of Work Among American Men and Women Workers, as Reported in 1960 and 2000 US Census

	Men		Women	
Hours Worked	1960	2000	1960	2000
1–29	7.1	8.7	18.4	19.4
30–39	7.4	6.9	18.9	16.4
40	47.0	45.2	46.4	46.1
41–48	20.5	12.4	11.9	7.8
> 48	18.1	26.8	4.3	10.3

services, the determination of "long" working hours varies across workers and types of jobs. However, suppose the threshold for determining a long working week is set at 48 hours. According to the US Census of Population for 2000, more than 25% of men and 10% of women usually work more than 48 hours a week.[2] See table 1.1; these numbers are noticeably larger than the corresponding percentages for 1960.[3] The spike in weekly hours at 40 as shown in table 1.1 for both 1960 and 2000, for both men and women, reflects the impact on employers of regulations requiring premium pay for covered workers after 40 hours a week.

2. The 2010 US Census was conducted shortly after the Great Recession, when labor markets had not yet recovered from the contraction. Hence, the 2000 Census is used to describe contemporary hours.

3. The question about weekly hours worked in the 1960 Census was not the same as that in the 2000 Census. In 1960, the question was "How many hours did this person work last week (at all jobs)?" In 2000, the question was "During the weeks worked in 1999, how many hours did this person usually work each week?" This change in wording may well have affected particular responses. The assumption is that, in comparisons of 2000 and 1960, the differences in responses induced by the wording is of second-order importance.

Introduction: Why Working Hours?

What are the characteristics of those men and women working these long hours? One way to answer this question is to use the 1960 and 2000 US Census data on individual workers, so as to define a dichotomous variable that takes the value of unity if the individual usually works 49 or more hours per week and of zero otherwise. Using least-squares, this variable is regressed on the individual's highest level of schooling attained, age, race, marital status, and self-employment (as well as occupation and industry fixed effects). The results are reported in table 1.2. In 2000, the probability of working long hours increases with the amount of schooling: the difference in the probability of working long hours between college graduates and high school graduates is 7.6 percentage points for men and 6.9 percentage points for women. This schooling gap is larger in 2000 than in 1960. The estimates are similar to those of Kuhn and Lozano (2008), who identify skilled salaried workers as those working longer hours in 2006 than in 1979.

Some of these patterns in working hours in America are evident also in British data. Using information from the 1998 Workplace Employee Relations Survey, Kodz et al. (2003) report the frequency distribution of usual weekly hours of market work (including overtime and any extra time) in Britain, shown in table 1.3. In both America and Britain, the distribution of hours exhibits a heap of observations at the point just before the hours at which mandatory premium rates of pay for overtime come into play.

When a logistic regression is fitted to describe the characteristics of those employees usually working more than 48 hours in Britain, those in well-paid occupations such as managers and professionals are well represented in the category of long hours, as are some operatives and assembly workers; one-third of employees aged in their thirties work more than 48 hours; and a

7

TABLE 1.2 Least-Square Estimates Attached to Variables Associated with Working Long Weekly Hours American Men and Women Workers in 1960 and 2000

Right-hand side Variable	Men		Women	
	1960	2000	1960	2000
Schooling				
High School Grad	0.022	0.029	0.001	0.006
Some College	0.006	0.048	0.002	0.022
College Grad	0.023	0.105	0.015	0.075
Age				
age 25–34 years	0.048	0.111	0.013	0.057
age 35–44 years	0.035	0.125	0.012	0.065
age 45–54 years	0.006	0.097	0.016	0.074
age ≥55 years	−0.032	0.039	0.018	0.053
Black	−0.064	−0.062	−0.022	−0.020
Married	0.086	0.060	−0.014	−0.029
Self-Employed	0.210	0.172	0.190	0.131
Nobs	320,840	632,256	159,518	586,718
Modv	0.181	0.268	0.043	0.103

Note: The left-hand side variable takes the value of unity if the individual reported usually working more than 48 hours a week and of zero otherwise. The reference Schooling category is "less than high school graduate" and the reference Age category is "less than 25 years." Black, Married, and Self-Employed are dichotomous variables taking the value of unity for black, married, and self-employed workers. Estimated standard errors are many times smaller than the estimated regression coefficients. These regression equations also include 24 fixed occupation effects and 15 industry fixed effects in 2000 and 10 occupation fixed effects and 11 industry fixed effects in 1960. Nobs means number of observations and modv means mean of dependent variable.

TABLE 1.3 Percent Distribution of Usual Weekly Hours of Work among British Working Men and Women in 1998

Hours	All	Men	Women
1–30	28	11	46
31–40	41	41	40
41–48	19	27	8
> 48	12	21	5

Source: From the 1998 Workplace Employee Relations Survey, as reported by Kodz et al. (2003).

higher proportion of men than women work long hours in labor markets.

Thus, in recent years, long working hours are a feature of between one-fifth and one-fourth of workers in America and Britain. How do these typical working hours in America and Britain compare with other relatively rich countries? The values given in table 1.4 were assembled by the Organization for Economic Co-operation and Development (OECD). As the table shows, the average annual hours of work per worker in 2015 in America are higher than in any other country listed, and British hours are higher than those of its neighbors, Germany and France.

Average hours of work have fallen since 1979 in all these countries, though they have fallen least in America. The longer hours in America reflect a larger fraction of workers working 40 or more hours per week and of more weeks worked in the year. In view of the few holidays mandated by statute, the United States has been

TABLE 1.4 Average Annual Hours Actually Worked per Person in Employment, 1979 and 2015

	1979	2015	Δ %
USA	1,829	1,790	−2.13
UK	1,813	1,674	−7.67
Australia	1,834	1,665	−9.21
Canada	1,841	1,706	−7.33
France	1,832	1,482	−19.11
Germany	2,186*	1,371	−37.46
Spain	1,954	1,691	−13.46
Japan	2,126	1,719	−19.14

*Germany's value for 1979 is for West Germany in 1983.

$$\Delta\% = 100\left[(H_{2015} - H_{1979})/H_{1979}\right]$$

Source: From OECD, Employment Outlook 2016, Table L.

termed the "no vacation nation."[4] Bick et al. (2016) present a careful analysis of working hours across countries that is attentive to differences in the manner in which the data are collected. Among other things, they find (consistent with the OECD values shown in table 1.4) that in a typical nonvacation week, American workers work longer hours than workers in the larger European economies.

Not only are there noticeable differences in working hours across countries, but also there are appreciable differences in

---

4. On statutory paid leave and public holidays across countries, see Ray and Schmitt (2007). On the distribution of usual weekly hours worked in 2015, see http://stats.oecd.org/. The incidence of workers working at night and on weekends is also higher in America than in Europe (Hamermesh and Stancanelli 2014).

Introduction: Why Working Hours?

hours across workers within countries. In the search for an explanation for these within-country differences, much has been written about the effect of differences in hourly wages across workers, but a great deal of that literature rests on uncertain foundations, an issue taken up in chapter 8. The emphasis in this monograph is on the *consequences* of such differences in working hours.

To provide a perspective on the subject, chapter 2 presents a brief history of the working hours for the typical American and British worker since the beginning of the nineteenth century. I am not a historian and I draw heavily on the research of others. The chapter concentrates on explanations that have been offered for the changes in these hours over time. Some of these explanations assume that working hours enter an enterprise's production function in special ways, and chapter 3 presents frameworks for investigating the empirical validity of these assumptions.

These frameworks for working hours in the production function are confronted with observations on marketable output and hours in chapter 4. Chapter 5 then describes research that has been conducted by others to help inform the nature of this output–hours relationship.

Chapter 6 considers the association between hours worked and a by-product of working hours: the health of workers. There is extensive literature on this relationship, but few nonspecialists appear to be aware of it.

In view of the findings presented in chapter 4 on the manner in which hours affect marketable output, chapter 7 then considers the copious research done by economists on the association between working hours and hourly earnings. The character of this research has changed over the last fifty years in unfortunate ways, with the consequence that much of it is of questionable value.

Finally, some conclusions are drawn in chapter 8.

It will become evident that the account that follows consists of organizing observations on working hours and the outcomes of these hours to draw inferences about the consequences of this relationship. At the same time, the topic of hours of work can be enriched by drawing on the personal experiences of workers themselves. Therefore, on a few occasions, I interrupt the formal analysis with brief descriptions of particular workers, reminding us that behind the dispassionate investigation of the effects of working hours are men and women for whom the time spent at work is an important component of their lives. These brief worker portrayalsprovide this material.

2

A Brief History of Working Hours

Without comprehensive surveys of the working population, it is difficult to make definitive statements about the typical length of work before the twentieth century, but some guarded inferences may be drawn from the research that has been undertaken.[1]

In Britain and America, before the Industrial Revolution, for those lacking inherited wealth, two activities dominated working life: agriculture and, especially for women, domestic service. In both cases, by today's standards the hours of labor were long—in agriculture, often from sunrise to sunset six or seven days a week, and in domestic service, sometimes longer. At the beginning of the nineteenth century, about 74% of American workers were employed in agriculture (and slave laborers constituted about 35% of all farm labor). In Britain in 1841, about one-fourth of workers were employed in agriculture. Note that at this time, most agricultural workers in Britain are believed to have been employees who often rented their dwellings from their employers. By contrast, in America, a larger fraction of agricultural workers (in

---

1. What follows on the history of hours in Britain draws heavily on Bienefeld (1972), Cross (1989), Hutchins and Harrison (1966), and Mitchell (1988); and in America on Cahill (1932), Lebergott (1964), Lester (1946), and Sundstrom (2006). On the nature of work before the Industrial Revolution, see Laslett (1984).

states free of slavery) were self-employed farmers working on the land they owned.

In America in 1832, women employed in domestic service represented 60% of all employed women, while in Britain in 1841, the corresponding number was 55%.[2] By the end of the nineteenth century, in both America and Britain, the total employment of men and women in domestic service had risen considerably. According to Lethbridge (2013, 9), "In 1900 domestic service was the single largest occupation [for women] in Edwardian Britain: of the four million women in the British workforce, a million and a half worked as servants, a majority of them as single-handed maids in small households" (9). Therefore, both agriculture and domestic service remained important sectors for employment into the twentieth century.

Describing the 1940s, Roberto Acuna started picking crops in California when he was eight years old. Rising at 4:00 a.m. on a weekday, he would work for several hours before going to school. After school, he would return to the fields for several hours of more work. On Saturdays and Sundays, he was in the fields from 4:30 a.m. until 7:30 p.m. (His account is contained in Terkel 1975, 30–38.)

As for domestic service, in America, Anderson (1936, 66) reports "In 1930, nearly one-third of all Negro workers—both men and women—were employed in domestic and personal service. The greater number of these workers—over 7,000,000 women and 110,000 men—had jobs as servants, largely in private homes." In 1933, Nannie Thompson of Alexandria, Virginia, reported (as contained in Sharpless [2010, 69]) her typical workday was 12 to 15 hours and longer "if there was a big function."

---

2. The American numbers come from Lebergott (1964, tables A-1, A-2, and A-12) and those for Britain from Mitchell (1988, table 2.B).

A Brief History of Working Hours

> *In Britain, an interesting case is provided by the experience of an American girl, Elizabeth (Lizzie) Banks (1894), who visited Victorian England to take on a series of jobs for "curiosity." One of these jobs was as a housemaid in the house of Mr. and Mrs. Allison. Lizzie's typical day was to rise at 6:00 a.m. to attend to her first task—that of shaking and brushing Mr. Allison's trousers that had been left hanging outside his room, then to brush Mrs. Allison's dress, then to take all the family's boots to the kitchen where they would be shined by the cook, and then to sweep and dust four flights of stairs, four halls, the study, and drawing rooms, and wake up Mr. and Mrs. Allison, providing each with a can of hot water. Then Lizzie was to eat breakfast. The rest of her day was specified by the hour until she may go to bed at 10:15 p.m. Lizzie went on to serve as housemaid for others and later she worked as a street sweeper, a flower seller, and a laundry girl. She wrote up her experiences and they were widely read in both Britain and America.*

In the eighteenth century, it was "customary" for workers in many trades in cities to work from "six to six"—that is, a 12-hour workday, from six o'clock in the morning until six o'clock in the evening. There were instances of longer hours and of occasions when workers in a trade combined to call for reductions in the length of the working day. Examples in Britain of such combinations were provided by building workers, tailors, masons, and bookbinders. These worker groups were not formal trade unions, but they are early examples of workers acting in concert to reduce their working hours.

Work was often conducted at home or in group living quarters, and workers were paid by the task completed or by the product delivered. In these instances, their working hours were unspecified by those giving them the tasks or purchasing their products. This domestic or putting-out system was displaced by large factories, where many people congregated to work under the supervision of a foreman and often with the aid of machines. There were complementarities among workers, so it was important to have workers together at the same time. Furthermore, because the machines were costly to run, it was wasteful to operate them when all workers were not present. In this way, the hours when a workday began and ended were specified, as were the days when attendance at the factory was required. There were exceptions, especially where payment-by-results constituted a component of earnings. In those instances, workers retained some latitude over when to quit work for the day. This tended to be the case for coal miners in both Britain and America.

Two developments reinforced the codification of working hours. One development was statutory legislation limiting the hours of certain workers—first, the hours of children and, then, of women workers. Adherence to the regulations required record-keeping of the working times; the absence of such records was prima facie evidence that the statutes were being ignored. The second development prompting the formalization of hours of work was the growth of organizations of workers that campaigned for clear specification of employment conditions, including working times. Working hours and days became scheduled or normal, even though employers retained the ability to alter the length of the workday if business conditions were not "normal." In this way, overtime was occasionally (or sometimes routinely) required of some workers, with in some cases premium rates of pay for hours of work beyond those deemed normal. When business was slack, workers might be placed on short time.

The formalization of hours and days of work was part of the development of "factory discipline," constraints placed by the employer on the employees at the workplace. Clark (1994) suggests that these constraints were devices to coerce workers to apply more effort to their jobs.[3] Notwithstanding this coercion, workers were willing to supply their labor to factories that applied these disciplinary methods, because the workers were paid more and, thereby, compensated for this control. Employers were willing to pay these wages because factory discipline raised worker effort and output.

According to Lester (1946), "working hours in English and American factories averaged about 13 a day and 75 a week around 1830" (344). After 1840, working hours declined in both America and Britain over the next hundred years. Around this trend, working hours moved with the business cycle, with longer hours in business expansions and shorter hours during recessions. Why did hours of work decline after 1840 in both countries? At least four classes of explanations have been offered: pressure on employers by trade unions; constraints of statutory legislation; initiatives by employers; and the effects of the invisible hand on labor markets. Each of these is considered in turn.

Trade Unions

From their formal beginnings in both America and Britain, trade unions agitated for shorter hours of work. In a number of instances, this has been the principal goal of their efforts, an objective in and of itself to reduce toil, drudgery, and stress. On occasion, a reduction in normal or usual hours of work has been an indirect way of engineering a rise in take-home pay. That is,

---

3. "In the nineteenth century, workers under factory discipline were dismissed, fined heavily, or locked out for the day for a whole variety of infractions. These included arriving a few minutes late in the morning, being absent from their machine, talking or eating at work, drinking beer, and whistling, singing, and engaging in other forms of horseplay" (Clark 1994, 131–132).

if actual hours worked remain unchanged and if longer hours command premium rates of pay, the reduction in usual hours results in a larger fraction of actual hours paid at overtime rates. In this way, reductions in standard or normal hours are an oblique means to higher weekly earnings.

Another circuitous purpose of reducing the usual hours of work was to generate an increase in employment. Although the number of workers may not be perfectly substitutable for average hours of work, they may be partly substitutable so that, in producing a given output, a cut in hours per worker could be partly offset by an increase in employment. If this increased employment came from the ranks of the unemployed, a reduction in hours worked would reduce unemployment.[4] The argument was summarized in a saying attributed to Samuel Gompers: "So long as there is one man who seeks employment and cannot find it, the hours of labor are too long." This notion has been popular for well over a century and sometimes goes under the term "work sharing." A number of governments have sought reductions in working hours with the goal of cutting unemployment; examples include the American federal government in the 1930s,[5] the French and German governments in recent decades, and some state governments in America that have tinkered with their unemployment insurance systems to encourage fewer hours of work in place of layoffs.[6]

---

4. "The fundamental causes of labor's demands for shorter hours have been the desire for leisure, and, more important still, the fear of unemployment" (Cahill 1932, 13).

5. Between 1929 and 1932, President Herbert Hoover urged employers to reduce the working week in the hope of reducing unemployment. In 1933, as part of the National Industrial Recovery Act, President Franklin D. Roosevelt created the National Recovery Administration, which encouraged firms to reduce working hours with the same aim.

6. In these states, instead of laying off workers, some of whom may be eligible for unemployment compensation, an employer can arrange for unemployment benefits to be paid to

A Brief History of Working Hours

In Britain in the first three or four decades of the nineteenth century, the decline of the domestic (or putting-out) system and the growth in factory employment are thought to have been accompanied by increases in the hours of work for many workers. This began to change in the late 1840s. Bienefeld (1972) attributes this change to the growing strength of craft trade unions. Examining individual industries in different places, Bienefeld documents a correlation between the strength of these unions and reductions in hours. In particular, the changes in hours in the years 1871 to 1874 were notable as to have "radically altered the country's conception of what constituted a day's work" (142). This success by these organized skilled workers in the engineering, iron and steel, pottery, printing, and other sectors prompted unskilled workers to unionize (the "new unionism").

In crediting the reductions in working hours to "the strength of trade unions," Bienefeld does not offer an index of their "strength," but refers to particular strikes (or the threat of such strikes) when reductions in hours were the primary issue at dispute and where reductions followed. There were instances of workers rejecting wage increases in lieu of hours reductions. When a major employer conceded to demands for reduced hours, other employers (in the same industry or location) were apt to follow. In this way, many of the workers in the metal trades and building industries saw their usual working week reduced from 60 to 54 hours in the first half of the 1870s. Some of these reductions were reversed in the recession that hit in the second half of the decade, suggesting perhaps that unionism may not have been the only explanation for cuts in hours or that the strength of unionism varied with the business cycle.

workers whose hours of work are reduced below their usual workweek. These schemes are known as "short-time compensation."

Hobsbawm (1964, 158–178) describes a remarkable episode when, after years of unsuccessful attempts at organization, British gas workers came together in 1889 to demand the replacement of 12-hour shifts with 8-hour shifts—demands that were coupled with pay increases for certain classes of workers. "Their demands were conceded virtually without a struggle," (158) although employers responded with the mechanization of a number of operations.

In a similar fashion, Cahill (1932) documents the strikes called by trade unions in America in the late nineteenth and early twentieth centuries that had the goal of reducing hours and that succeeded. In these cases, a working day of 10 hours was reduced to 9 or 8 hours. In some instances, these reductions in hours turned out to be transitory and, a few years later, the 10-hour day was reestablished. This return to a 10-hour day sometimes occurred in states whose legislatures had passed bills mandating an 8-hour day! In those instances, the unions claimed that their demands were designed simply to enforce an existing law.

Lester (1946, 346–347) divides the American manufacturing industry into a unionized sector and a non-unionized sector in the years between 1890 and 1925, and reports average full-time weekly hours in each sector. He is well aware that this union–nonunion differential in hours does not control for other factors that may differ between these two branches of manufacturing. Nevertheless, the differences are suggestive: the proportionate union–nonunion hours differentials are −13.8% in 1890, −15.5% in 1900, −17.2% in 1910, −15.7% in 1920, and −13.4% in 1925; these differences (in absolute magnitude) not far from union–nonunion earnings differentials.

Lester also presents the hours of union and nonunion workers in US construction in 1936, and again these indicate longer working hours for nonunion workers. Brown (1959) makes similar union–nonunion comparisons in Britain and finds that "by the end of the century, indeed, the [union] organized trades were more consistently marked off from the rest by their shorter hours than by any other difference" (79).[7]

Bienefeld (1972) claims that a distinctive pattern in the hours reductions in Britain points to a role for trade unions: "while wages could be raised under the normal competitive pressures of the market, this was not true of the reduction of hours. It was because of this difference that institutional pressure was deemed essential if reductions in hours were to be achieved" (143). He draws attention to the fact that, whereas increases in wages were continuous and incremental, reductions in the usual hours of work were discontinuous and substantial. Thus, after 1840, while wages tended to rise in small amounts almost year by year, hours did not decline in a continuous decremental fashion but in irregular, discrete steps with little change in the years between.

Bienefeld identifies merely four periods (1872–74, 1919–20, 1946–49, and 1960–66) in a period of 150 years from 1820, when reductions in hours were considerable and were experienced by a large number of workers in Britain. Similarly, Matthews et al. (1982) write, "Reductions in hours of work did not take place at a steady rate but were concentrated, with ratchet effects, in phases when the bargaining power of labor was strong: in the early 1870s, in 1919, in 1948–49, and in the postwar period" (511).

---

7. Eichengreen (1987) relates differences in the logarithm of summer and winter daily hours worked among 3,334 Iowa men in 1894 to a union membership dichotomous variable and to other characteristics of these men, and concludes "there is little evidence that they [union members] had . . . success in reducing the length of the workday" (514). Unfortunately, he does not present estimates of the union–nonunion differential in hours that do not hold constant the occupational differences in hours. If the union effect on hours extended to all workers in an occupation (union and nonunion members alike), then the occupational differences in hours that he measures are a surrogate for union differences.

There is also some suggestion of this pattern in American manufacturing: whereas the typical workweek of production workers "inched downward" (Whaples 1990, 393) from about 1870 to 1910, in the next decade the workweek fell by six hours. In the American iron and steel sector, average weekly hours remained almost constant from the 1890s to 1919, and then in the postwar contraction, they fell "dramatically" to 1924 (Shiells 1990, 380). Millis and Montgomery (1938, 465) refer to this feature in American industry: "One characteristic of the standard working day or week is that its length does not change gradually within an industry, as do wages. On the contrary, hours manifest a tendency to shift from one plateau to another and ordinarily each new standard prevails for a period of years." Writing in 1984, Hunnicutt states "the shorter hour movement reached some sort of historical plateau nearly 40 years ago" (375). Much has been written about the "rigidity" of money wages and yet such rigidity appears to be no less apparent in hours of work.

Why have the reductions in working hours taken this discontinuous form? Hicks (1963, 107) suggests that, because of complementarities in production among workers (i.e., it is often critical for production that employees work at the same time), reductions in hours must generally be granted by an employer to all his workers. By contrast, an employer may raise wages for some but not all workers; for instance, some workers may see attractive job opportunities at another firm and, to induce them not to leave, the employer might raise their wages without extending this increase to all workers. It may be some time before such differential treatment becomes known and sufficient resentment about the inequity induces the employer to pay all workers the same wages. The firm's wage increase for all workers is thereby distributed over time rather than being concentrated at a particular moment.

A similar process for working hours—with some workers enjoying fewer hours of work than remaining workers—may not be desirable, even if feasible, with the result that workers usually start and end work at the same time and work the same number of hours in a given workplace. In addition, an employer may be uncertain of the consequences of reductions in hours, and in the event of undesirable results, might find it difficult to restore the original hours of work. This explanation places employers in a central role in determining the hours of work, but they are under pressure from organizations of workers.

There is a literature investigating the effects of trade unions on hours of work that uses contemporary data. For instance, as a by-product of his research into union–nonunion wage differentials, Lewis (1986, 106) reports twenty-one estimates of union–nonunion differentials in hours of work. These relate to American workers in the 1960s and 1970s, and they describe weekly hours, annual hours, and annual weeks worked. In sixteen of these estimates, holding constant a number of other variables, hours of work are lower among union workers than nonunion workers. American unions reduced both the incidence and the extent of overtime hours (Trejo 1993).

Comparing cities of like size in the 1960s, Ashenfelter (1971) consistently finds a negative association between weekly hours of work and the incidence of unionism for some public workers. The difference ranges from about −5% in smaller cities to −9% in larger cities. Using information from the US Census Bureau/ US Bureau of Labor Statistics' Current Population Surveys since the 1960s, Frandsen (2016) concludes that unionism reduces the length of the workweek by a small but statistically significant amount. In Britain, after the legislative campaign against trade unions in the 1980s and early 1990s, and the decline of unionism in Britain, "there is only weak evidence to suggest that British unions were having an impact on working time in the 1990s and 2000s" (White and Bryson 2016, 9).

Therefore, there is both historical and contemporary evidence of union workers enjoying shorter working hours than nonunion workers. There is also evidence that, when unions are weak, there is little or no union-nonunion difference in hours. From this cross-sectional pattern in various years, it is plausible to infer that the activities of trade unions contributed to the decline in working hours in both America and Britain for a hundred years, beginning in 1840. When only a relatively small share of workers are covered by union-negotiated agreements, to account for the reduction in hours among many workers, a mechanism is needed to explain how these union-negotiated agreements are extended to nonunion workers.

That mechanism exists when workers have employment opportunities at more than one workplace: if one employer concedes to trade union pressure and grants his employees shorter hours, other employers in the market may anticipate greater difficulties in retaining and recruiting workers, insofar as workers prefer employment with the employers offering shorter hours. To forestall these turnover costs, those other employers are induced to follow the lead of the first-moving employer and adopt shorter hours schedules. In this way, familiar market forces transmit the union-secured gains of shorter hours in one firm to workers in many other firms. Indeed, Bienefeld (1972) observes that employers "always preferred, if reductions in hours had to be granted, that all of them should be affected alike" (113). Explicit collusion among employers was not necessary though if one were to accept Adam Smith's view, collusion may well have been present in many cases.[8]

---

8. Smith (1776, book 1, chap. 8) wrote "Masters [employers] are always and everywhere in a sort of tacit, but constant and uniform combination.... To violate this combination is every

Employers

The typical employer in America and Britain resisted pressures to cut hours and, at times, has tolerated strikes and opprobrium to oppose such reductions. Even when the favorable experiences of employers who reduced their employees' hours were presented to them, other employers in the same industries or regions found reasons to discount these experiences.

Britain

An example is provided by the case of the Salford Ironworks in Manchester, Lancashire. In 1892, with 1,200 employees working 53 weekly hours, William Mather, the employer, proposed a 48-hour working week without altering wages and piece rates. The union (the Amalgamated Society of Engineers) agreed to the proposal, with the understanding that there would be a reversion to the previous schedules if the reduction in hours was unsatisfactory to either party. In fact, after implementing the scheme for a year, neither party wanted to return to a 53-hour workweek: with the reduced hours, production was slightly higher, operating costs (on lighting, machinery maintenance, and electricity) were lower, and piecework earnings were slightly lower. The 48-hour workweek became permanent at the factory.[9]

---

where a most unpopular action, and a sort of reproach to a master among his neighbours and equals. We seldom, indeed, hear of this combination, because it is the usual, and one may say, the natural state of things which nobody ever hears of."

9. See Mather (1894) for his description of this experiment and for his replies to skeptics and critics.

This experience caused the British government to cut hours at its ordnance factories. Hours of work at the Woolwich Arsenal were reduced from 9 to 8 hours a day without any reduction in output, an experience that induced Ernst Abbé to experiment with reductions in hours between 1899 and 1901 at the Zeiss Optical Works at Jena, in Germany. From his meticulous record of this experience, Abbé determined that the performance of "eight hours of work exceeded nine hours work."[10]

Most private employers dismissed Mather's experience. The Federation of Engineering Employers cautioned other firms not to follow Mather's example, claiming that if 48 hours became the norm for the industry, "it would result in a very large increase in the cost of production, would enable our foreign competitors to secure the control of the market, and therefore would bring ruin to many firms" (McIvor 1987, 728). Indeed, a few years later, the federation organized a lockout to resist the union's calls for a 48-hour workweek.

Mather's findings resemble Robert Owen's (1816) claims almost seventy years earlier. At a time when hours of work for children toiling in textile mills averaged around 13 hours a day, Owen reported that, when hours in his New Lanark mills were cut from 11 3/4 each day to 10 hours, production increased and there was not "the smallest alteration in any of the other circumstances" of production (such as machinery and raw materials). According to Owen, the reduction in working hours induced a "greater desire of the individuals to perform their duty conscientiously . . . than when they are forced to do their duty" (Hutchins and Harrison 1966, 22–23). The response by British Members of Parliament

---

10. The quote is from Goldmark (1912, 161), who provides a careful description of Abbé's work and of other early investigations into changes in working hours.

to Owen's statements displays the same disbelief that Mather's fellow employers exhibited in the 1890s.

America

In America in 1914, when Henry Ford more than doubled the daily pay of his workers at his motor car company, he also cut their daily hours from 9 to 8. (The working week remained at six days until 1926, when it was reduced to five days.) Why did Henry Ford reduce the working hours? There are several reasons. First, Ford conducted a survey of workers to determine the reasons for the firm's high labor turnover. At the top of the list of the "chief causes of dissatisfaction and unrest among employees" was "too long hours" (Meyer 1981, 1060). In addition, Ford disliked trade unions, and as unions were organizing successfully in 1913, his 1914 actions constituted a classic response of a non-union employer to discourage his workforce from becoming unionized: improve the well-being of workers to reduce or eliminate the appeal to organize.[11]

After these reductions in hours and days of work, Ford's labor turnover dropped and his labor productivity increased. However, this increase in productivity cannot be ascribed simply to the cut in hours, as Ford made other changes in the operation of his plant at that time. Nevertheless, Ford's actions in 1914 and

---

11. On the growth of unions at that time, according to Wolman (1924, table 3, p. 34), total union membership grew by 34.5% between 1909 and 1913. Moreover, the IWW had undertaken campaigns in Ohio in 1913 and announced their goal to unionize the automobile workers. So, there was ample concern for an employer such as Ford to entertain the possibility of union organization. Indeed, according to one of Ford's biographers, "there was a growing threat of unionization" (Curcio 2013, 77). On the threat effect of unionism and its correlates, see Lewis (1963, 23–27).

in 1926 were met with an "outburst of criticism" (Cahill (1932, 252) from other employers.

An unusual and interesting case is that of W. K. (Will Keith) Kellogg who, in November 1930, introduced a 6-hour working day for most of his employees at his breakfast-cereal plant in Battle Creek, Michigan. He did so on the advice of the company's president, Lewis J. Brown, an English immigrant much influenced by the practices of William Hesketh Lever, who had built a new factory with an adjacent housing estate (Port Sunlight) for his employees and who campaigned in Britain in the 1920s for a standard working day of 6 hours. The stated purpose of Kellogg's 6-hour day was to increase employment by replacing three 8-hour shifts with four 6-hour shifts. It lasted until the Second World War, when wartime demands and the shortage of workers led to a reinstatement of the 8-hour day. After the war, the 8-hour day became a contested issue, and was not applied to all workers until 1984, when a plant-wide policy of 8 hours was accepted by the union; see Hunnicutt (1996) for more on this.

William Mather, Ernst Abbé, Robert Owen, Henry Ford, W. K. Kellogg, and William Lever were unusual among employers in taking the initiative and implementing shorter hours for their workers. Most employers and employer organizations resisted workers' agitation for reductions in hours. As Cahill notes (1932, 27), "the reduction in hours by pioneer employers has been less common than that brought about by legislative or trade union action." Employers also spent resources in opposing trade union calls for shorter hours and legislation drawn up to mandate ceilings on working hours. In Britain, writing of the 1919–1939 period, Lowe (1982, 266) describes the employer position on proposals to reduce working hours as one of "a blanket opposition to all change even when, in the late 1920s, such change was shown to be to their possible commercial advantage." In America, trade union pressures to reduce hours "were met by exceptionally strong business opposition" (Hunnicutt 1984, 378).

Employers and Their Beliefs

"Religion has been one of the forces at work in the establishment of the five-day week" (Cahill 1932, 253). The textile mills that introduced (without legal or trade union compulsion) a five-day working week in America were those owned by Orthodox Jews, whose Jewish employees sought to observe the Sabbath on Saturday. Kellogg was a member of the Seventh-day Adventist Church, for whom Saturday, not Sunday, is the Sabbath. Hunnicutt (1996, 21fn.72) has conjectured that the problems in trying to avoid work on Saturdays in a society where most people treated Sunday as the day of rest may lie behind Kellogg's concern with hours of work. The inspiration behind Kellogg's shorter workday was William Lever who, in his beliefs, was a Nonconformist in England who "practised the social gospel of Christianity in his own life" (Bradley 1987, 191).

In Britain, in the vanguard of benevolent Victorian capitalists (introducing shorter working weeks, longer holidays, and other employee benefits) were Quaker employers such as George and Richard Cadbury, Joseph and Seebohm Rowntree, and Cyrus and James Clark. These employers placed a value on their ethical dealings (including shorter hours) with their employees. For instance, at a time when the typical workday was 10 hours and when a few firms had granted a 9-hour day, George and Richard Cadbury set an 8½-hour day (less 1/2 hour for breakfast for which coffee was provided) and 20 minutes for lunch (Williams 1931, 52).

In 1919, Seebohm Rowntree set down a 44-hour working week, but allowed the employees to decide by means of a ballot whether the 44 hours were to be worked over five days or in five-and-a-half days. They chose the former, although management would have preferred the latter. See Briggs (1961, 234). These Quaker employers managed successful and popular companies that commanded a devoted following among consumers. The companies did not find it impossible to combine a shorter working week with commercial success.

Why, therefore, were their employment practices not imitated more than they were? The notion that workers might work more effectively if they worked fewer hours, so that a given reduction in the number of work hours might not translate into the same reduction in the effective work done, was rarely entertained. In fact, several observers have inferred from what employers have said or written that employers assumed that an x% reduction in hours worked would imply an x% reduction in output. This belief will be named here the assumption that there are *unit returns to hours of work*.

Statutory Legislation

Statutory legislation has been advanced as an explanation for the reduction in working hours. In both America and Britain, legislatures introduced constraints on the hours that employers might ask of their workers.

Britain

In Britain, regulations on hours of work took the form of a series of Factory Acts, the earliest in 1802. The 1802 Act was directed toward the length of the workday of apprentices, many of them children, in spinning and weaving workshops. Although some

A Brief History of Working Hours

children worked alongside a parent in the mill, in the late eighteenth century most child workers in these mills are believed to have been orphans and apprenticed to employers. Eager to transfer responsibility for these orphaned children to someone else, parish officers placed the children with employers notwithstanding the little schooling the children received as mill workers. Occasional surveys suggest the children often worked as much as 13 or 14 hours a day. Six days of work each week were common, with shorter hours on Saturday. In addition, their working conditions were unhealthy and dangerous.[12] The Health and Morals of Apprentices Act in 1802 limited the working hours of the apprentices to 12 a day. Night work was to be eliminated within two years of the Act's passage. The Act applied to all apprentices of any age in these mills, but not to those child workers who were not apprenticed. Subsequent legislation extended the initial restrictions on hours worked to all children, and in 1833, an enforcement agency with factory inspectors was created. One Inspector's Report claimed that "the output of eleven hours' work might be greater than that of twelve" (Hutchins and Harrison 1966, 122).

In 1847, an Act to Limit the Hours of Labour of Young Persons and Females in Factories was passed, which capped a working day for children at 10 hours in the textile mills on Monday to Friday and at 8 hours on Saturday—a working week of 58 hours. Two years later, this was amended to a working week of 60 hours, which is believed to be the modal working week in British manufacturing at the time. In some workplaces (textile mills among them), the work done by children was

---

12. From data collected a century later, by which time things should have improved, the fiber-laden hot air caused cotton workers to have a mortality rate from respiratory diseases that was two and a half times that of agricultural workers. See Wohl (1983, 277).

complementary to that of adults, so restrictions on the working hours of children tended to put downward pressure on the hours of adults also.

Debates arose over the desirability of legislation restricting working hours. In these debates, employers tended to argue against them, claiming that these reduced hours would harm British industry in competition with foreign firms. Economists of the Classical school tended to approve of restrictions placed on child labor from a humanitarian perspective, but they were far less sympathetic to proposals to regulate the hours of adults, especially those of men.[13] Writing of the Classical economists, Blaug (1997, 210) concludes, "All in all, the friends of factory legislation [that restricted hours] were not far wrong in regarding 'political economy' as a major obstacle to factory reform."

For many nineteenth-century economists, fundamental principles of laissez faire were at issue: adult workers ought to be free to ask for any working hours they wish and employers ought to be free to ask for any working hours they want. Because parents could not always be relied upon to protect the well-being of their children, because some children were orphans, and because children were not "free agents," legislation with respect to the working hours and schooling of children might be excepted from these laissez faire principles.

There were differences of opinion over whether legislation on hours ought to apply to women (Robbins (1978, 102). Rarely, it seems, did these economists present careful reasoning that provided the logic for the purported damage wrought on a firm by restricting its hours of work. Rather, an overarching faith in

---

13. Blaug (1958) and Nyland (1989) provide accounts of economists' views.

unfettered laissez faire accounted for most economists' reactions to constraints on hours of work.[14]

Typical of the dire consequences that many economists claimed would follow from shorter hours legislation, Henry Fawcett (1872), professor of political economy at the University of Cambridge, wrote that

> If successful in an eight hours' agitation, an agitation might commence in favour of fixing the day's work at seven or even at six hours. If, however, such restrictions [on hours] were imposed, it can scarcely be doubted that industry would be placed in so unfavourable a position that it would be hopeless for England to attempt to compete with foreign countries. It might thus happen that not only her foreign trade would be sacrificed, but she would be undersold in her own markets. It is not too much to say that her commercial prosperity would cease, and that a fatal blight would be thrown upon her industry. (121)

No model or representation of the effect of mandatory reductions of hours on the production unit accompanied these dire predictions.

For the most part, notwithstanding pressure on Parliament to regulate the length of working hours,[15] statutory legislation in Britain on hours since the Factory Acts is noticeable by its absence. Hours of work were expected to be set by collective bargaining between a union and an employer, or by a joint

---

14. John Ramsay McCulloch may have been one of the few economists who provided some reasoning in the debate over statutory restrictions on hours. Maintaining that child labor and adult labor tended to be substitutes in production, he supported restrictions on the hours of child labor on the grounds that the reduction in supply would put upward pressure on the wages of adults; see Sorenson (1952).

15. A well-argued example is Webb and Cox (1891).

management–labor committee such as a Whitley Council. For example, among the issues presented to the Whitley Council for the Administrative and Legal departments of the Civil Service in 1919 was "the attempt by the Treasury to force on its general classes an eight—instead of a seven—hour day" (MacRae-Gibson 1922, 5). What legislation there was applied to a small number of workers (such as children) or to particular industries, retailing being a clear example. Thus the Shop Hours Regulation Act in 1886 limited the hours of work of shop workers under the age of eighteen years to less than 74 weekly hours. In 1911, the Shops Act required shop workers to have a half-day holiday each week. This became known as "early closing day."

This noninvolved stance for statutory legislation changed in 1998 when, after much debate, Britain adopted the European Union's Working Time Directive, which set a limit on weekly hours. At the time of writing (2017), the regulations state that an employee may not work more than 48 hours per week when averaged over seventeen weeks. The worker may opt out of this limit, although the employer cannot require the worker to opt out. Opting out is not possible for workers in certain industries, principally those involving transport, so as to avoid driving when drowsy. Opting out is not a permanent arrangement, and it can be canceled when the employee so chooses. Professional workers, managers, and the self-employed are presumed to set their own hours and are exempt from the 48-hour maximum. The regulations also stipulate that most workers are entitled to paid leave and rest periods.

Five years after the Working Time regulations came into effect, an analysis of its operations indicated that individual opt-outs were "widespread . . . because of the ease with which the [48-hour] limit can be avoided" (Barnard et al. 2003, 223). More recently, the thorough review by Devlin and Shirvani (2014) for the UK Department for Business Innovation and Skills suggests "tentatively" some effects toward a shorter working week, though they appear to be minor. Those working long hours tend to be for men in skilled occupations, a finding also reported by Kodz et al. (2003) and reported in chapter 1. All in all, statutory legislation of working hours in Britain over the last century appears to have affected relatively few workers.

America: State Regulations

In America, in the textile factories at the end of the eighteenth century, working hours were from 12 to 15 a day, depending on the season, as in Britain. Children constituted a significant fraction of the workforce in these mills. A larger proportion of the workers in America were women than in Britain at the same time. With laws in Connecticut in 1813 and in New Jersey in 1816, the first labor legislation in America was directed toward the schooling of children. Subsequently, between 1842 and 1853, restrictions on the working hours of children were passed in a number of states, although as little or no enforcement mechanism was provided, these laws tended to be disregarded.

In the 1860s, some states—Massachusetts, Connecticut, New York, Illinois, Missouri, and Wisconsin—enacted legislation specifying a legal maximum of 8 hours a day of work in the absence of a collective agreement. Cahill (1932, 96) describes these statutes as "futile" because many exceptions were specified and few or no resources were devoted to their implementation. They did lead to strikes by some labor unions to induce employers to apply the regulations, but their success was mixed. Budgets for factory inspectors were enacted in the 1880s together with regulations limiting hours of children to 10 per day. Minimum ages for factory work were raised to twelve or thirteen years.

Some state and local governments passed statutes regulating the working hours of particular groups of workers, such as government employees, and in some instances, for the employees of private companies that contracted with state or local governments. Again, they were of little consequence, and in some instances, the judiciary took a hostile stance to them. Some states identified women employees as needing some protection in part because women tended to be outside the fledgling unions.

In 1874, Massachusetts passed a maximum 10-hour workday (and 60-hour week) for women and for those aged less than eighteen years. Problems of enforcement were addressed in subsequent amendments. By 1911, the Massachusetts law specified a ceiling of 54 weekly hours. In some states, attempts at legislation for shorter hours for women did not survive judicial challenges until California's law setting a limit of an 8-hour day and 48-hour week for women in a large number of jobs was upheld both by California's Supreme Court and by the US Supreme Court in 1912. However, in 1914, when California and Washington state proposed broad 8-hour laws that applied to both women and men, they were soundly rejected by the electorates.

State legislatures tended to be more sympathetic to restrictions on the working hours of women, and the supporters of such restrictions hoped that, if successful, these regulations might subsequently be extended to men. Hence, much of the effort by advocates of constraints on hours at the state level was directed toward female workers. Women's health was believed to be particularly susceptible to fatigue that accompanied long hours. By the 1920s, most states had passed regulations governing the hours of women workers. (Summary statements of these laws are provided by Jones 1975.) Using data from the Census of Manufactures in various years, and comparing industries covered by the provisions of legislation with those not covered, both

Jones (1975) and Goldin (1988) investigated the link between changes in hours and the timing of the legislation, comparing states with legislation to states without maximum hours legislation, and contrasting the hours of men with women, using both actual hours and scheduled hours. They conclude that the contribution of this legislation to the decline in hours was "minimal."

America: Federal Regulations

President Martin Van Buren's executive order in 1840 mandating no more than a 10-hour working day for federal government employees is usually identified as the first federal attempt at a minimum standard. Then, for the same class of employees, in 1868, the US Congress passed an 8-hour law. While it was in the federal government's purview to define the working hours of its own employees, it was not clear the government had the constitutional authority to specify hours for employees of businesses that contracted with the federal government.

Congress passed an act in 1912 requiring all US government contracts to contain a provision specifying an 8-hour day constraint with nontrivial monetary fines for infractions and requiring government inspectors to check that the requirement was satisfied. Even this act allowed for exceptions, including presidential overrule in the event or imminence of war, an exception invoked within a few years of its passage.

Interstate commerce provided a constitutional role for the federal government in regulating working hours, and clearly the railroads were involved in interstate transport. In 1907, the Hours of Service Act established the principle of federal management of the hours of railroad employees (endorsed by the Supreme Court in 1911) and the Adamson Act of 1916 stipulated 8 hours as the "measure of standard of a day's work" for railroad employees, with additional pay for overtime. This overtime wage premium was also used in the Walsh-Healey Public Contracts Act of 1936 and in the Fair Labor Standards Act (FLSA) of 1938, both of which stipulated that workers covered by the terms of the Act receive time-and-a-half pay for each hour worked beyond a specific level.

The FLSA set that level at 44 weekly hours in October 1938, at 42 hours in October 1939, and at 40 hours in October 1940. Using information on variations in the extent of coverage of the Fair Labor Standards Act (for instance, when first introduced, most employees in wholesale trade were covered, whereas few in retail trade were covered until years later), Costa (2000b) estimates that the FLSA had a substantial effect in reducing hours of work in the covered sector, the consequence both of the minimum wage provisions and of the overtime provisions of the Act.

This overview of legislation in America from the early nineteenth century until the legislation of the 1930s does not give much support to the notion that statute law plays a major role in reducing the hours worked by a large part of the labor force. Laws were passed without much in the way of enforcement, and trade unions had good cause to complain that employers were simply ignoring the statutes. At the same time, the mass of individuals working at the hours beyond which overtime pay premia apply is prima facie evidence of the empirical relevance of such statutory (or trade union–negotiated) rules in affecting the distribution of hours since the late 1930s.

A Market Explanation

During the period when hours of daily work for the typical worker fell from 12 hours to 10 hours, and then to 8 hours, and when seven days of work gave way to six and later to five days in a typical week, the real earnings (hourly, weekly, and annually) rose for many of those working in labor markets. Therefore, with rising earnings and less time at work, the typical worker's material standard of living rose substantially. When hours fell, it was often at a time when real earnings were rising noticeably. Bienefeld (1972, 40) draws attention to the negative association over time, at least until 1820, between hours of work and wages in Britain, both in the factories and "where the hours of work were entirely at the discretion of the working man." In the nineteenth century, movements in real hourly earnings followed an upward trend while hours of work trended downward.

When such a clear association exists, it is tempting to identify one variable as the "cause" and the other as the "effect." Thus, in discussing trade unions, mention has already been made of how a cut in hours might result in a rise in average hourly earnings, brought about by increasing the fraction of working hours paid at overtime rates. When this occurred, the reductions in hours were an indirect path to or a "cause" of higher hourly earnings.

Another line of reasoning suggests that the reductions in hours were induced by rises in pay. This was the approach of Lewis (1957) who, using market demand and supply curves for hours of work, endowed these curves with special properties. For the supply of hours, the key assumption is that "tastes for leisure [by workers] are very stable in the long run," by which he meant that the supply curve of hours of work (reflecting the income-work preferences of workers) was essentially unchanged for decades. Regarding the demand curve, he assumed "to a first approximation" that, at the market wage, the employer cares about the aggregate hours worked by the firm's labor force but not about each worker's choice of hours. Consequently, the typical employee works the hours he chooses.[16]

Diagrammatically, as shown in figure 1, when the market hourly wage is w_0, the employer's demand curve for hours per worker is the horizontal line D_0. This demand curve shifts upward over time (to D_1, D_2, D_3) as incomes and expenditures grow and the derived demand for labor rises and wages increase. With the assumption that the supply curve is unchanged (because "tastes for leisure are very stable"), this implies the typical worker's labor supply curve is negatively sloped; the worker responds to higher wages by working fewer hours. Hours of work fall when real wages rise, signifying the income effect exceeds the substitution effect in the typical worker's preferences.

A horizontal demand curve for hours per worker of the sort shown in figure 1 is implied by a production function in which hours of work enter the production function with a unit elasticity—that is, a production function in which an x% change in hours changes output by the same x%. A production function with this property of *unit returns to hours* will have a marginal product of hours that is independent of hours, as pictured by each demand curve in figure 1.

Expressed differently, the demand for hours that Lewis posited as a "first approximation" is one in which the law of diminishing returns to hours does not apply. If Lewis had postulated the law of diminishing returns, his demand curve for hours per worker would have been sloping downward to the right. In this case, an

---

16. Lewis suggests that years of extensive unemployment or meaningful statutory legislation (such as the Adamson Act in the railroad industry and the Fair Labor Standards Act) might also affect hours. In his opinion, trade unions may well have reduced hours in some industries or occupations but they have been less influential for the economy as whole because unionism covered such a small proportion of total employment.

A Brief History of Working Hours

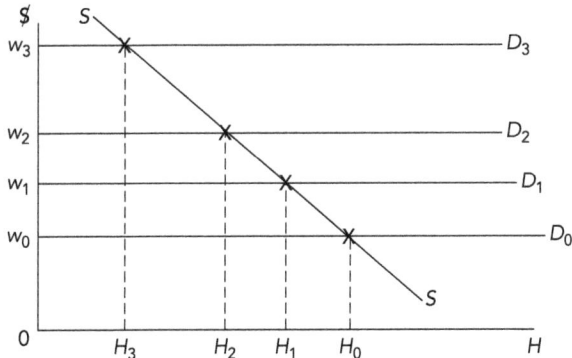

FIGURE 1 Lewis's Explanation for the Decline in Weekly Hours Worked.

Note: In the graph above, the real wages of a typical worker are measured on the vertical axis and the worker's hours of work on the horizontal axis. The **X** combinations of real wages and work hours trace out the effect of a horizontal labor (hours of work) demand function rising over time (D_0, D_1, D_2, D_3 above) and an unchanged negatively-sloped labor (hours of work) supply function (given by S S in the figure). Over time, higher real wages (w_0, w_1, w_2, w_3 above) are associated with shorter hours of work (H_0, H_1, H_2, H_3).

exogenous rise in wages would have induced employers to reduce hours. Thus, an alternative market explanation for the negative association between hours and wages over time is that employers reduce the use of a more costly input—hours of work—when hourly earnings rise. This alternative explanation requires the demand for hours per worker to be downward sloping with respect to wages.

Lewis's assumption of the typical employer's indifference to the hours worked of the typical employee might be viewed as flatly contradicted by the vigorous opposition that most employers displayed toward calls for shorter hours. Indeed, about ten years later, Lewis (1969) proposed a different characterization of the determination of working hours, one used by Rosen (1974) in his model of hedonics. Lewis's 1957 model merits interest because of its remarkable durability in the economics profession.[17]

For instance, while acknowledging its shortcomings, Lewis's 1957 model is used to organize Huberman and Minns's (2007) thinking about movements in weekly hours across countries since 1870. Research in economics has paid and still pays considerable attention to the link, if any, between the hours of market work of individuals and their average hourly earnings, and this will be taken up more fully in chapter 7.

America and Britain Contrasted

Although working hours tended to decline after 1840 in both America and Britain, in most years the hours worked tended to be lower in Britain. For example, in the mid-1840s, "New England mill women were required to work at least five hours more each week (seventy-four at Lowell) than their English counterparts" (Merish 2017, 55). By 1870, the average annual hours of full-time production workers in America exceeded those in Britain by 340 hours, a consequence both of fewer days worked per year and a shorter working week in Britain (Huberman and Minns 2007). Indeed, in American manufacturing in 1880, Atack and Bateman (1992) report that a majority of employers stated their typical working day was 10 hours. By contrast, the previous decade (the 1870s) in Britain was a time of agitation to reduce daily hours to 9

---

17. Boppart and Krusell (2016) have released a long working paper that offers an explanation for the decline in working hours over time that is identical to Lewis's explanation in its key features—workers' preferences in which the income effect of higher wages exceeds the substitution effect and employers' preferences that play no essential role—without any reference to Lewis or to the influence Lewis has had on the subject.

(or 54 weekly hours), and in a number of instances, the agitation was successful.

Not only were working hours for the typical worker longer in America than in Britain by 1880, but also Wright (2010, 234) suggests that the intensity with which the typical American worker worked was greater, and he cites the statements of immigrants testifying to the difference. Smuts (1953) also provides declarations to this effect by foreign observers, but after providing other impressions, Smuts writes "[t]he American worked long hours (slightly longer than in England) . . . yet he spends less physical energy than the European because he was provided with every facility to lighten and speed his task. . . . [T]here are reasons to doubt that Americans worked harder than other people" (17–18). In other words, the typical American worker had the benefit of greater complementary inputs. If work effort were easier to measure, these contrasting claims would be easier to adjudicate.

On the eve of the First World War, in many cases in both America and Britain, the seven-day workweek had been replaced by a six-day workweek, with shorter hours worked on Saturdays. But whereas annual days of vacations and national (public) holidays increased in Britain from fourteen in 1870, to twenty in 1900, and to thirty in 1938, in America, these days rose from four in 1870, to five in 1900, and to seventeen in 1938 (Huberman and Minns 2007). Why were (and still are; see table 1.4, in chapter 1) working hours longer in America?

In both countries, principles of laissez faire were strong and legislatures were reluctant to impose constraints on employers. A key difference between the two countries concerned the extent of and the public respectability for labor unions. In both America and Britain, the end of the nineteenth century and the first decade of the twentieth century were years of growing unionism. In America, starting from a low base, union membership in 1913 was six times its membership in 1897, while in Britain union membership in 1913 was about two and a half times its membership in 1897. This growth is manifest in the autobiographies of workers of that time.

Paul Evett, an English printing compositor, describes a highly contentious strike in 1911 brought by the Typographical Association against a printing works in Newport, Monmouthshire, that hired a new worker (a woman) at below union-negotiated rates. Even though the firm took on strikebreakers, "It took a long time, but eventually the firm came back into the fold," writes Evett. (In Burnett 1974, 337–338.)

In America, Rose Schneiderman (1905), an immigrant from Poland, worked 10 hours a day as a cap-maker on piece rates in New York. Dissatisfied with their employer, Rose and other women determined that "we girls needed an organization. The men had organized already, and had gained some advantages, but the bosses had lost nothing, as they took it out on us." The women sought help from the National Board of United Cloth and Hat Makers. The subsequent "big strike" in which other unions extended their help resulted in a "glorious victory" for the workers. She reports that many saleswomen on "the East Side work from 8.00 a.m. till 9.00 p.m. week days and one-half day on Sundays ... so they certainly need organization."

While unionism was growing at this time in both countries, the extent of unionism in America was lower than that in Britain: the fraction of the UK's labor force who were members of trade unions in 1901 was about 10.8%, whereas the corresponding fraction in America was approximately 2%.[18] Moreover, in Britain, by the end of the nineteenth century, unions were gaining in

---

18. The UK numbers are drawn from Mitchell (1988, tables 4 and 13). According to Troy and Sheflin (1985, table 3.41), American trade union membership in 1897 was 455,00 (the 1901 figure is not provided) and the US labor force in 1900 was 29,073,233 (from Historical Statistics of the United States, US Census, Series D 1–10).

general acceptance and in status. Some employers saw benefits in a workforce organized by dependable union men. As Phelps Brown (1983, 19) expresses it, at this time "[i]n the experience of Victorian employers it was the most skilled, responsible, and steady workmen who took the lead in the unions and unionism had an uplifting influence on the weaker brethren. Industrial relations were found to be at their best when strongly organized unions entered into voluntary negotiations with stable associations of employers." There was a growing tendency in Britain to recognize unions as legitimate expressions of the interests of workers.

Evidence of this attitude is provided in the passage of the Trade Disputes Act in 1906 in Britain, which granted unions immunity from damages sought by an employer in a strike. British trade unions accepted the political framework and were deeply involved in the formation of the Independent Labour Party (ILP) in 1893 and in the Labour Representation Committee in 1900. The 1906 General Election resulted in fifty-three Labour Members of Parliament, a number of them trade union officials.

The significance of unions was furthered in Britain by the role that a number of union officials and union members played in Nonconformist (such as Methodist, Presbyterian, Baptist, and Congregational) places of worship. Ensor (1936, 528–529) writes of the late nineteenth and early twentieth-century unionism, "One way or another, the rising labour movement owed an immense debt to nonconformity. The fund of unselfish idealism which maintained the early I.L.P. came mostly from this source; and the methods whereby its branches were run and financed were borrowed directly by its members from their experience in religious organizations." Scotland (1997) observes that, as early as 1836, the speaker at a Methodist rally was calling for an 8-hour workday: "Methodism made two very prominent and overt contributions [to the labour movement]. These were in the provision of leadership and in offering models of organisation" (48). It is no accident, then, that in some British unions (such as in the printing industry), the local organizations are still called "chapels." To modify slightly the line associated with Harold Wilson (1964, 1), British trade unions owed more to Methodism than to Marx.

As an example, consider George Edwards (1922), who was born in 1850 and died in 1933. After a year in a workhouse as a child, his first job at six years old was to scare crows away from fields recently seeded. The job required him to be up "soon after sunrise and remain in the fields till after sunset." As an adult, he became an agricultural laborer in the county of Norfolk, working "from five in the morning to five in the afternoon." He writes: "the labourer barely ever sees his children by daylight except on Sunday" (17–19). After his wife taught him how to read, he became a lay preacher in Primitive Methodist chapels. Subsequently, he was active in organizing the precarious agricultural unions. He stood in several general elections until he became a Labour Party Member of Parliament in 1922. In 1930, he was knighted "for services to agricultural workers."

At the time that, in Britain, fifty-three representatives of the labor movement entered Parliament, in America, the Industrial Workers of the World (IWW) was formed at a convention in 1905. The IWW was a radical organization with anarchist and syndicalist sympathies that firmly rejected the notion of working within the political system. Earlier, the American Federation of Labor (AFL) advocated the improvement of workers' conditions through actions at their places of work, not through a political organization such as a formal commitment to a political party. The IWW struggled to find acceptance or, even, tolerance. While the tight labor markets of the First World War brought a jump in the membership of the AFL, the IWW's opposition to the war brought it disgrace and legal suppression.

American unions were largely detached from the political system until the Second World War, when they offered a no-strike pledge for the duration of the conflict (as they had in the First World War); in return, they were closely involved in the decisions of the War Labor Board. After the war, concern about communist influence within unions kept the unions removed from close involvement in the political process, notwithstanding their financial support for the Democratic Party. In contrast, in Britain in the postwar period, trade union officials were brought into government and, in some instances, into the Cabinet.

Trade unions played a greater role in British public life not because of an overarching statutory framework that defined and protected them (such as America's National Labor Relations Act) but because there was a series of indirect measures (supported by representatives of unions in Parliament) designed to bolster collective bargaining and unionism, such as requiring government contractors to accept union-negotiated wages and working conditions and setting minimum wages (by Work Councils) in sectors that might compete with unionized industries (Pencavel 2004). Britain's nationalized industries were legally required to recognize and negotiate with trade unions. Broadly, whereas American unions often found themselves on the defensive and struggling for wide approval, British trade unions were accepting of and accepted by the established political system, and the unions used that political system to support and extend themselves.

This system changed in the 1980s when, in response to damaging strikes against public-sector monopolies, Margaret Thatcher's government introduced legislation that withdrew much of the indirect support. Thus, at least until recent years, differences in the stature, influence, and legacy of labor unions are the principal explanation for working hours in Britain being less than those in America. Both inside and outside government, British unions have routinely campaigned with positive effect for better employment conditions for workers.[19]

What Has Been Learned?

The chapter's discussion references several hypotheses that have been advanced to account for the decline in working hours during the nineteenth century and into the twentieth century and for the preferences of workers for shorter hours to be realized. These hypotheses differ in the mechanism by which the reductions in hours were achieved. One points to the influence of trade unions, another to the effects of statutory restrictions, and a third proposes that amenable employers allowed their employees to work less when their hourly earnings rose. Each of these explanations suggests that workers took some of their real wage increases in the form of shorter hours, rather than in the form of greater consumption of goods and services.

These mechanisms were not independent of one another, however. Thus, Cahill (1932, 26–27) maintains that, between 1914 and 1932, the average working hours of unionized workers in America tended to be 4 to 6 hours per week fewer than those of nonunion workers. She cautions that this understates "the full measure of the effectiveness of trade union activity since the hours of unorganized workers are not independent of those

---

19. Recently, British unions have spoken against zero-hour contracts. These are essentially on-call employment arrangements whereby the employer does not specify any working hours for the employee but requires the employee to be available for work only when the employer needs him or her. The employer pays only for hours actually worked, so the worker's earnings are unpredictable.

established by the organized, insofar as the latter set up a standard toward which legislative enactment tends and the unorganized struggle." These interdependencies frustrate the identification of a single explanation for the decline in working hours between 1840 and 1940. When, for instance, the effect of trade unions on working hours depends upon other factors—such as the posture of the law on hours or on the willingness of employers to experiment with shorter hours—the search for a single explanation for the reduction in hours during these years is bound to be strained.

The pattern of the reductions—the discontinuous cuts in hours for a few years with long intervening periods without change, and the contrast with the continuous rise in hourly earnings—confirms Hicks's judgment that nonmarket forces were the principal explanation for the reductions in hours. Hicks (1963) suggests that market forces tend to be more consequential in raising wages and that "long hours worked in the early days of the Industrial Revolution were reduced, it is well known, mainly by State regulation and Trade Union action" (106).

State regulation played a role in the early nineteenth century in Britain with respect to child labor, but reductions in the working hours of adults were more likely to have been the product of trade union pressure. Statutory legislation in America and, in particular, the requirement that employers pay time and a half for hours worked beyond 40 is the ostensible explanation for the mass of workers in recent years who report they usually work 40 hours a week.

This brief account of the economic history of working hours has identified several enduring beliefs about the way in which hours of work figure in production. One is the notion that a given proportional change in the number of hours worked results in the *same* proportional change in output. In his analysis of the Classical economic reasoning about the effects of mandatory reductions in working hours, Blaug (1958, 217) refers to this as "an essential feature of the classical analysis." Employers also took this as given: "employer resistance to shorter hours has postulated a loss of output fully proportional to the reduction in hours" (Denison (1962, 39); employers believed that a mandatory reduction in hours of work will "decrease output in exact proportion to the reduction in hours" (Cahill 1932, 15). This belief is referred to here as the hypothesis of *unit returns to hours*.[20] As mentioned earlier, this assumption is contained in Lewis's 1957 explanation for the decline in working hours.

The relationship between proportional changes in hours and proportional changes in output is called the *elasticity of output with respect to hours*, $\mathscr{E}(X, H)$, a unit-free measure of the response of output to changes in hours. For small changes or differences, this may be expressed in the (natural) logarithmic equation:

$$\Delta ln(X) = \pi \Delta ln(H)$$

where X denotes output, H is hours, and Δ stands for change. This equation may be derived from the production function

$$X = AH^\pi$$

where A stands for all inputs other than hours that are held fixed in the change equation. The hypothesis of *unit* returns to hours is such that, in the preceding equation, the parameter π is unity. If π is unity, the average product of hours (X/H) and the marginal product of hours $\partial X / \partial H$ are equal to A, a constant, and the law of diminishing returns does not apply to hours of work.

---

20. "Returns to scale" relates to the effect on output of the same proportionate change in all inputs. This is not to be confused with "returns to hours" here, which holds all inputs fixed other than hours.

A Brief History of Working Hours

A more general hypothesis is that of *constant returns to hours*, when the previous equation describes the output–hours relation, but π need not be unity. With constant returns to hours, the effect of a proportional change in hours worked on the proportional change in output (the elasticity of output with respect to hours) is the same, irrespective of the number of hours being worked.

This hypothesis is a common assumption in research on production functions and is embodied in the production function popularized by Charles Cobb and Paul Douglas (at least when the labor input is measured such that hours worked per worker are distinguished from the number of workers).

The purpose of the research reported in the next chapter is to present a framework that allows for examination of these hypotheses and, more generally, to determine how hours of work enter a firm's production function, with particular attention to the effects of long working hours. After this framework is developed, it will be applied to several sets of observations in chapter 4.

3

Conceptual Framework

The goal in this chapter is to specify production functions in which working hours affect output or performance in a more flexible manner than has been typical of production functions to date. The production functions specified will be applied to sets of observations that permit an assessment of whether the Classical economists, Lewis in 1957, and many employers were correct in maintaining the hypothesis of *unit returns to hours*. If they were correct, the law of diminishing returns does not apply to working hours. More generally, is the hypothesis of *constant returns to hours* an appropriate description of the manner in which working hours enter the production function? The reason that the hypothesis may not be an apt description arises from the fact that the services of labor consist of a number of different dimensions.

Clearly, one dimension is *time*: an employee supplies his time to an employer. In addition, as Coase (1937), Simon (1951), and others have recognized, the typical employer–employee relationship is one in which the employee accepts the directions of the employer or the employer's representatives. But the employee retains an ability to execute these directions with varying degrees of diligence and devotion to the work. In so doing, the worker supplies his energy, enterprise, and initiative to cooperate with

Conceptual Framework

management and with other workers in teamwork. This bundle of characteristics or qualities will be labeled "effort." Typically, at the start of work, the employee has a stock of effort that is spent over the day and over the week at work. By the end of the work period, this stock of effort has diminished. The extent to which his effort runs down will depend upon the nature and length of the work, on the attributes of the worker, and on the environment in which the individual works.

With this general framework, suppose that in a single workplace in a given week, a group of workers produce an output, the amount of which is given by X. This output depends on the *effective* hours of these workers, where "effective hours" means not simply the number of nominal hours these individuals work but also the effort per hour that these individuals put into their work. Suppose L denotes these effective hours and suppose the quantity of other factors (such as the number of workers, the plant and equipment, and raw materials) contributing to output are given as Z. Then assume the link between output and effective hours and other inputs, the production function, may be written as follows:

(1) $$X = f(L) \cdot g(Z)$$

where f stands for the nature of the dependence of X on L. The manner in which factors other than effective hours relate to output is given by $g(Z)$, and this is assumed to take the form of kZ^{η} where η is a parameter. Assume that the effective hours of these workers are the product of their nominal hours per worker H and the effort per hour E applied by these workers to their work is:

(2) $$L = E \cdot H.$$

Effort is not directly observed, but it may be inferred from the values of variables that are observed. If effort is unobserved, *effective* hours, L, are not observed. Only *nominal* hours, H, are observed.

Work effort or the intensity of work—Alfred Marshall's (1920, 438) term—has long been recognized as a dimension of the input of labor. Reder (1999, 182) goes so far as to identify work effort as the defining feature of the labor input writing, "What distinguishes labor from other productive services is the worker's ability to vary productivity by altering effort." Hicks (1963, 2) wrote "labour is a two-dimensional quantity, depending both on the number of labourers available, and upon their 'efficiency'— the amount of labour each is able and willing to provide." If there are two dimensions to the input of labor, as Hicks maintains, there is need not only for an analysis of a worker's supply of hours of work but also of a worker's supply of effort at work.[1]

There have been brave attempts to construct indicators of work effort. An example is Green's (2001) use of workers' answers to survey questions asking about the effort they apply to their jobs and the stress accompanying their work. The index of work effort he constructs for British workers in the 1980s and into the 1990s rises over calendar time with women workers displaying a larger rise than men.

The fact that effort is unobserved causes Bowles (2004) to declare that labor contracts are essentially incomplete. At the same time, a literature has emerged that describes the design of compensation schemes that induce agents (the workers) to behave in a manner that benefits the principal (the employer). See Salanié (2005, esp. chap. 5) for an exposition. Under certain conditions

---

1. On this, see Dickinson (1999), Lin (2003), and Pencavel (1977). Becker (1977) also has some characteristic insights.

(including the ease of observing output), piece-rate methods for paying workers are deemed particularly conducive to the promotion of work effort. If this argument is valid, the application of payment-by-results methods (of which piece rates are an example) should dispose workers to produce greater output when other inputs (including working hours) are constant. There are a number of empirical studies that have confirmed this.

For instance, examining a company making windows for cars, Lazear (2000) draws upon greater worker effort to account for a large increase in output (measured by the number of glass units installed per 8-hour day) that followed a switch from time rates of pay to piece-rate methods. Part of this increase in output was attributed to the addition of more able workers who were attracted by piece-rate work to the firm. Remarkably, the firm's profits also rose, implying that during years when the firm was paying its workers by the hour, it was not maximizing profits.

A longitudinal study (Paarsch and Shearer 1999) of workers of a tree-planting firm in British Columbia uses the payroll records on the daily output of eighty-nine planters over a five-month period in 1994 to examine the effect of variations in the piece rate on work effort. The firm varied the piece rate to reflect differences in the type of terrain on which the planters worked—that is, piece rates were higher where planting conditions were more exacting. Taking account of the firm's choice of piece rate and allowing for different abilities among the workers, the authors estimate an elasticity of work effort with respect to the piece rate of 2.14, implying that effort was highly responsive to changes in the piece rate.

Treble (2003) undertakes a similar analysis of coal miners in County Durham in the 1890s. He identifies a period when changes in the piece rate (tied by a sliding scale to the price of coal) induced changes in output per shift per fortnight, and yet hours of work were unchanged. Treble estimates an elasticity of work effort with respect to the piece rate of 1.07.

The calories consumed by each individual at work, conditional on his or her body size, are used by Foster and Rosenzweig (1994) as an indicator of the work effort expended by each of almost one thousand agricultural workers in the Philippines. During the period examined, these individuals were compensated on some occasions by piece rates and on other occasions by time rates. Holding constant variables measuring the characteristics of work and workers, a typical worker's daily consumption of calories was 23% greater when working on piece rates than on time rates, suggesting greater effort was applied when working on piece rates.

A different type of evidence comes from Roger Kaufman's (1984) interviews of British employers during a sharp business contraction. Asking why these employers were reluctant to cut wages, Kaufman reported

> Employers invariably felt that work effort was endogenous and depended on worker motivation and satisfaction. As a result, they believed that there was substantial potential variation in both the quantity and quality of output for a constant input of labor hours. Because many of the smaller employers did not have foremen or supervisors and often had to leave the "shop," they relied heavily on the goodwill of their employees. (107)

Using observations on the threshing of grains and the reaping of wheat (measured as bushels per worker day), Gregory Clark (1987) showed that agricultural productivity in the nineteenth century in the northern states of America and in Britain was twice that in continental Europe. After entertaining various possible explanations, Clark attributes the difference in productivity to work effort or to the intensity of work: "it was differences in the rate at which simple manual tasks were performed which largely account for the differences in output per worker across countries in the early nineteenth century and over time" (4250).

All these and other empirical investigations invoke work effort to account for changes and differences in observed outcomes. There is ample evidence for work effort to be treated as a relevant variable in the analysis of labor markets. None of these studies, however, links work effort with working hours, as is proposed here.

At this juncture, two different lines of reasoning will be followed to address the fact that effort is unobserved and yet is hypothesized to vary with working hours. Both of these approaches postulate the form of the production function given by equation (1), the definition of effective hours of work given by equation (2), and the specification of the role of other inputs as $g(Z) = kZ^\eta$. They differ in their assumptions about the form of $f(L)$ in equation (1) (that is, in the manner in which effective hours, L, enter the production function) and they differ in how the unobserved variable, effort per hour, E, varies with nominal hours of work, H.

An Expression Inspired by Charles Cobb and Paul Douglas

One approach writes $f(L)$ as equal to L and assumes that the unobserved effort per hour depends on the number of hours worked in the following way:

(3) $$E = (H)^\theta.$$

The parameter θ converts hours into effort such that if θ is negative, each additional hour worked results in a reduction in effort per hour; if θ is positive but less than unity, effort increases with hours, but the increase in effort diminishes as hours lengthen. If θ is unity, effort is equivalent to hours; and if θ is zero, effort is independent of hours. Substituting equation (3) into equation (2) yields $L = (H)^{1+\theta}$ and, given the assumption that $f(L) = L$, equation (1), the production function, becomes

(4) $$X = (H)^{1+\theta} \cdot kZ^{\eta}.$$

To this point, a single group of workers has been assumed to produce an output in a specific week. Suppose information is available on output, hours, and other inputs for a number of weeks and for other groups of workers. If equation (4) describes these weeks and these workers, then using j to index each group of workers and t to index each week, equation (4) may be written as

$$X_{jt} = (H_{jt})^{1+\theta} \cdot k_{jt} Z_{jt}^{\eta}$$

where, evidently, θ and η are assumed to be the same across workers and over time.

Suppose k_{jt} varies across weeks and workers in the following fashion:

$$k_{jt} = exp(\beta_0 + \omega_j + \varepsilon_{jt})$$

where exp is the natural exponential function, β_0 is a constant, ω_j is a fixed effect for each group of workers, and ε_{jt} embodies random variations over workers and over time in the association between other (nonlabor) inputs and output. Then, expressing the previous equation in natural logarithms results in

(CD-1) $$lnX_{jt} = \beta_0 + \beta_1 lnH_{jt} + \eta lnZ_{jt} + \omega_j + \varepsilon_{jt}$$

Conceptual Framework

where $\beta_1 = (1+\theta)$. Equation (CD-1) is a familiar Cobb-Douglas production function in which the logarithm of output is linear in the logarithm of the inputs and, approximately, proportional changes or differences in output are linearly related to proportional changes or differences in inputs. The parameter β_1 has a particular interpretation because θ may be inferred by subtracting unity from β_1: if β_1 is greater than unity, as Feldstein (1967) speculated, then θ in equation (3) is positive and effort per hour increases with hours worked; if β_1 is unity, θ is zero and effort is independent of hours; if β_1 lies between zero and unity, hours of work are subject to the law of diminishing returns, θ is negative, and as hours lengthen, so effort per hour declines.

β_1 is the point elasticity of output with respect to hours—what was called $\mathscr{E}(X, H)$ earlier. Hence, equation (CD-1) provides a simple way of examining the hypothesis of unit returns to hours: as mentioned, the Classical economists assumed that β_1 is unity, as did Lewis (1957) in his explanation for the decline in working hours in the nineteenth and early twentieth centuries. If β_1 is unity, the marginal product of hours equals the average product of hours, and both are independent of hours worked. It is a special case.

Occasionally, (CD-1) is estimated by subtracting the logarithm of one input from both sides of the equation. For instance, in his study of production functions in US manufacturing industry, Griliches (1967) subtracted the logarithm of worker-hours from both sides of equation (CD-1). If $\ln H_{jt}$ is subtracted from both sides of (CD-1), an expression for output per hour is derived:

(CD-1)# $\quad ln(X/H)_{jt} = \beta_0 + \rho ln H_{jt} + \eta ln Z_{jt} + \omega_j + \varepsilon_{jt}$

where $\rho = \beta_1 - 1$. Changes in output per hour worked are sometimes called changes in labor productivity. If β_1 is less than unity and effort per hour declines with hours, then ρ will be negative and increases in working hours will lower labor productivity measured this way.

Although the production function given by equation (CD-1) is familiar in broad outline, it differs from Cobb and Douglas's production function, which is often estimated in several ways. When it was introduced and estimated, the input of labor was measured without any consideration of hours of work; it was measured entirely by the number of employed workers. Cobb and Douglas (1928) were aware of the neglect of hours of work and addressed this later. Usually, in subsequent research, the labor input is measured by forming the product of the number employed and hours worked per worker. This assumes that the elasticity of output with respect to hours per worker is the same as the elasticity of output with respect to the number of workers, a constraint that Feldstein (1967) questioned.

The favorable empirical performance of Cobb and Douglas's production function has been demonstrated many times, but it has some undesirable features in a study examining how an input—hours—enters a production function. By construction, for any continuous production function, the elasticity of output with respect to an input (here, hours of work) equals the marginal product of that input divided by the average product of that input.[2] Because the elasticity of output with respect to hours $\mathscr{E}(X, H)$ equals β_1 in equation (CD-1), Cobb and Douglas's function

---

2. The elasticity of output with respect to hours, $\mathscr{E}(X, H)$, may be written as $(\partial lnX)/(\partial lnH)$ or $(\partial X/\partial H)/(X/H) = MPH/APH$, where MPH is the marginal product of hours and APH is the average product of hours. The hours at which the average product of hours is maximized are those at which $\mathscr{E}(X, H)$ is unity. This is at $H^\wedge$ hours in figure 2.

restricts the marginal product of hours MPH to be a fixed proportion of the average product of hours, APH regardless of the level of hours. This restriction may be innocuous or even appropriate over some ranges of hours (such as those hours that an informed and purposive owner-manager would choose), but this restriction conflicts with the usual characterization of how the marginal and average products of an input change as more of that input is used.

The figure drawn routinely in textbooks has the appearance of figure 2, in which, at hours less than $H^\wedge$, the marginal product exceeds the average product and, at hours greater than $H^\wedge$, the marginal product of hours is less than the average product of hours. In figure 2, the marginal product of hours is not a constant proportion of the average product of hours, so that the Cobb and Douglas function (CD–1) cannot reproduce the curves drawn in figure 2. Equation (CD–1) restricts the elasticity of output with respect to hours to be the same at all hours—whether workers are working 30 hours a week or 60 hours a week.

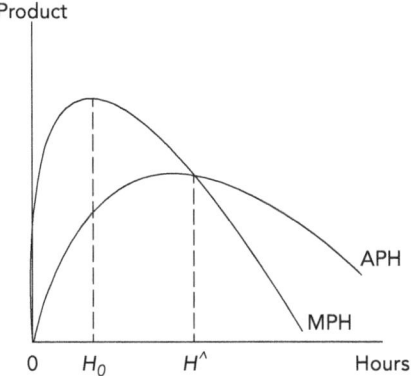

FIGURE 2 The marginal product of hours (MPH) and the average product of hours (APH) as implied by production functions usually drawn in textbooks.

This feature may be addressed by allowing β_1 to vary with hours of work—that is, write β_1 as β_{1jt} and retaining the logarithmic form, suppose $\beta_{1jt} = \beta_2 + \beta_3 lnH_{jt}$, in which case equation (CD-1) becomes

(CD-2) $\quad lnX_{jt} = \beta_0 + \beta_2 lnH_{jt} + \beta_3 (lnH_{jt})^2 + \eta lnZ_{jt} + \omega_j + \varepsilon_{jt}.$

Whereas in equation (CD-1), the elasticity of output with respect to hours, $\mathscr{E}(X, H)$ is a constant, β_1, in equation (CD-2) a point estimate of $\mathscr{E}(X, H)$ is $\beta_2 + 2\beta_3 ln(H_{jt})$ and $\mathscr{E}(X, H)$ varies with hours unless β_3 is zero.[3] Because $\mathscr{E}(X, H)$ equals the ratio of the marginal product of hours to the average product of hours, if the textbook relation pictured in figure 2 obtains, $\mathscr{E}(X, H)$ will be greater than unity at relatively short hours and $\mathscr{E}(X, H)$ will be less than unity at relatively long hours. After the marginal product of hours peaks at H_0 hours in figure 2, $\mathscr{E}(X, H)$ will decline as hours increase.

Note that, if lnH_{jt} is subtracted from both sides of equation (CD-2), if $\beta_2 < 1$, and if $\beta_3 < 0$, the resulting expression for output per hour or labor productivity will decline with hours at a greater rate than implied by equation (CD-1)*.

Equations (CD-1) and (CD-2) will constitute the basis for one set of estimating equations. Equation (CD-1) will be called the Cobb and Douglas conventional specification and equation (CD-2) will be the Cobb and Douglas augmented specification.

An Expression Inspired by Benjamin Gompertz

Let's return to a single workplace in a given week where a group of workers produces an output, X. A second approach to specifying

---

3. There are other ways to allow the elasticity of output with respect to an input to vary with that input. For instance, Anxo and Bigsten (1989) specify a translog production function, and they fit this to observations on Swedish manufacturing industry.

Conceptual Framework

the manner in which hours of work enter the production function retains equation (2) but posits that $f(L)$ in equation (1) is the natural exponential function:

(5) $$f(L) = exp(L) = exp(E \cdot H).$$

Assume also that workers' hourly effort varies with the length of the working day or week or year in the following manner:

(6) $$E = \alpha_1 + \alpha_2 H$$

where $\alpha_1 > 0$ and $\alpha_2 \leq 0$. Combining equation (6) with equation (2),

$$L = E \cdot H = (\alpha_1 + \alpha_2 H)H = \alpha_1 H + \alpha_2 H^2$$

and, if $\alpha_2 < 0$, effective hours, L, are a positive concave function of nominal hours of work, H.

Substituting the previous equation into equation (5) and the result into equation (1) yields

(7) $$X = f(L) \cdot g(Z) = \left[exp(\alpha_1 H + \alpha_2 H^2) \right] \cdot kZ^\eta$$

where, again, $g(Z) = kZ^\eta$. Equation (7) describes a single group of workers in one week. If it applies to other groups of workers, and if it applies to many weeks, this may be recognized by modifying the equation to allow for different workers j and other weeks t as follows:

(8) $$X_{jt} = \left[exp(\alpha_1 H_{jt} + \alpha_2 H_{jt}^2) \right] \cdot k_{jt} Z_{jt}^\eta$$

Taking natural logarithms of equation (8) and assuming $k_{jt} = exp(\alpha_0 + \omega_j + \varepsilon_{jt})$, we arrive at equation (GZ–2):

(GZ–2) $\quad lnX_{jt} = \alpha_0 + \alpha_1 H_{jt} + \alpha_2 (H_{jt})^2 + \eta lnZ_{jt} + \omega_j + \varepsilon_{jt}.$

Equation (GZ-2) relates proportionate differences in output to absolute differences in hours of work, to proportionate differences in other inputs, to a fixed effect for each group of workers ω_j, and to an additive stochastic term ε_{jt}. The effect of *changes* in hours on output depends on hours worked. The estimation of equation (GZ-2) will provide estimates of α_1 and α_2, which allow computation of both the effort–hours relation, as equation (6), and the expression for effective hours. If effort per hour does not decline with hours worked, a_2 is zero in equation (6). In this case, equation (GZ-2) may be written:

(GZ-1) $lnX_{jt} = \alpha_0 + \alpha_1 H_{jt} + \eta lnZ_{jt} + \omega_j + \varepsilon_{jt}$

Equation (GZ-1) will be named the Gompertz specification and equation (GZ-2) as the Gompertz augmented specification. These specifications hark back to Benjamin Gompertz's (1825) work on the relation between mortality rates and age.

The type of relationship in equations (GZ-1) and (GZ-2), in which logarithmic differences in one variable (in this case, output) depend upon absolute differences in another (hours), is familiar in empirical economics. H. Gregg Lewis (1963) used this specification to measure the impact of trade unions on relative wages. Jacob Mincer (1974) related differences in the logarithm of earnings across workers to differences in their years of schooling and their years of labor market experience, where experience takes a quadratic form as do hours in equation (GZ-2).

The equations inspired by Cobb and Douglas's double-logarithmic formulation and by Gompertz's semi-logarithmic formulation allow for (but do not require) the increase in output to decline as hours lengthen. The increase in output declines as hours lengthen if β_1 is positive but less than unity in equation

Conceptual Framework

(CD-1), and this occurs in equation (GZ-2) if α_2 is negative. Why might β_1 be less than unity and why might α_2 be negative?

The answer is that there is considerable evidence that, after a point, an individual's performance in an activity declines as time spent (without interruption) in that activity lengthens, a circumstance sometimes referred to as fatigue, stress, tedium, or inattention. Performance may be indicated by output, but other indicators have been used to describe the decline in performance as time lengthens: for example, the reaction time of truck drivers, the incidence of spoiled products, the frequency of accidents, and the absenteeism or illness rate of workers. These indicators suggest that performance is not a constant proportion of working time. Grandjean (1969, 74) presents graphs that map various indicators of "output" against working hours, and these show output rising with hours but at a decreasing rate, the precise shape varying with the nature of the work.

The development of these equations—(CD-1), (CD-2), (GZ-1), and (GZ-2)—allows the examination of various hypotheses regarding the effect of working hours on output. Though they are similar, the use of two different forms to express the relation between output and working hours reduces the possibility that the results are happenstance, or that they pertain only to one particular specification of the relation. Each expression facilitates the examination of different issues in the role of hours in production. Cobb and Douglas's expression is amenable to the measurement of the elasticity of output with respect to hours. It is also a familiar form to economists and has extensive applications. Gompertz's expression provides a flexible form to examine nonlinear effects of hours in production. Workers, employers, and legislators express the duration of work in terms of hours and days of work (not the logarithms of hours or days), and this is exactly how hours are represented in Gompertz's functional form.

Both specifications rely on the distinction between nominal hours of work, H, and effective hours of work, L. When the marginal product of hours MPH is computed, this shows the effect on output of a small increase in nominal hours. The magnitude of this effect will depend on how nominal hours translate into effective hours, a translation that depends on work effort. Note that, because effort is not observed, the hypothesized links between nominal hours and work effort in equations (3) and (6) are not unique; positive linear transformations of these equations will yield the same form of the production functions.

Agents care about output, not its logarithm. Although there may be reason to express output in terms of its logarithm in order to compute the parameters of the production function, there is also a case to be made for converting differences in the logarithm of output into differences in output. Therefore, once the values of the production function parameters have been determined, the implied logarithm of output will be calculated for each specified hour of work. Forming the anti-logarithm of the implied logarithm of output yields implied output. In this procedure, other inputs will be held fixed at their arithmetic mean levels and stochastic elements will be set to their mean. When least-squares is used to estimate the parameters, the mean value of the computed residual is, of course, zero. It is instructive to graph the resulting output–hours association. The parameters of the specified equations—(CD-1), (CD-2), (GZ-1), and (GZ-2)—thus are estimated in the next chapter by conventional least-squares. If they were needed, plausible instrumental variables are not available.

In relating output to hours worked in this way, both the behavior of workers and the behavior of the employer or plant

Conceptual Framework

manager are being revealed. In the typical plant, the employer sets the working schedule, the time when workers start and finish work, and the days the plant is open. The work schedule responds to the demand for the plant's output and expresses the employer's understanding of the hours that employees need to produce the required output. That output will depend on the effort that workers apply to their tasks. Their effort will depend on a number of variables, including the length of the working week or working day, as well as the pecuniary returns to expending effort. Therefore, the plant's production function expresses the conduct of both the employer and of the employees.

4

Estimates of Production Functions

Consider now the observations to which the production functions presented in chapter 3—equations (CD-1), (CD-2), (GZ-1), and (GZ-2)—will be fitted. All observations are provided in the book's data appendix. They are designed to describe the input–output technology of a single plant or of a group of like plants. They do not aggregate simply or conveniently across workplaces with different technologies. Therefore, with one exception (that using observations collected by Kossoris and Kohler 1948), the applications that follow estimate these equations to a single plant or to a group of similar plants.

Observations Collected by Horace Vernon During the First World War

The Setting

After the first encounters in 1914 with a much larger and better prepared army, it became evident that Britain's "contemptible little army" (Kaiser Wilhelm II's words) needed to be bigger and better equipped. A campaign to induce young men to enlist in the military was the response to the first problem. Beginning in 1916, volunteers were augmented with men conscripted to fight. As these young men responded to the call to sign up for the military, young women answered the appeal to replace the young men in the workplace and especially in the factories producing the war materiel.[1] Conventions and rules restricting work effort and output were abrogated for the duration of the war, and employers reinstated Sunday work.

Notwithstanding these developments, a shell shortage was alleged and, to alleviate the purported shortage, women, older men, and youths were called upon to work very long hours in the munitions factories. Some questioned the wisdom of such long hours: Did they have damaging health consequences for the workers, which in turn would hinder production? Therefore, after the reorganization of H. H. Asquith's administration in May 1915 and David Lloyd George was appointed to head a new Ministry of Munitions, the Health of Munition Workers Committee (HMWC) was set up "to consider and advise on questions of industrial fatigue, hours of labour, and other matters affecting the personal health and physical efficiency of workers in munitions factories and workshops."

The committee's chair was Sir George Newman, who had already written an influential study on the causes of infant mortality and who later became the chief medical officer to the Ministry of Health. The committee commissioned a number of studies and surveys on the health of these workers and the inferences drawn from these surveys were published in the committee's memoranda.[2] In turn, these were reported in the United States by

---

1. In July 1918, there were well over four times the number of women employed in the munitions industries as there were in July 1914 (Marwick 1970, 91).

2. For an examination of the recommendations of the Munition Workers Committee, see Pencavel (2015b).

the US Department of Labor in its bulletins[3] and by the National Industrial Conference Board (1917). An important dimension of the committee's investigations turned on the association between worker output and their hours of work.

On this, particularly careful investigations were undertaken by Dr. Horace Vernon, a physiologist and Fellow of Magdalen College, Oxford. One contemporary scholar writes of the research of Vernon and other investigators for the HMWC in glowing terms: "these studies have rarely been surpassed in terms of scientific method and attention to detail, and remain one of the most important data sources in the field" (Spurgeon et al. 1997, 372). Vernon spent time at a munitions factory in the Midlands, from which he gathered information about the factory's operations on four groups of workers. The workers were on piece rates, and payroll records contained information on their output. The workers were unaware that their hours and output were being investigated.

In collecting information on the workers' weekly hours of work, Vernon did not rely on what workers claimed their hours to be, nor on what the employer reported. Because these workers were operating with machines powered by electricity, he used watt meters to determine when the workers were actually engaged in work. The starting time at work was determined by the moment watt meters at workstations indicated the supply of electricity had reached a level to run the machines. (The use of watt meters followed Abbé's methods at the Zeiss Optical Works a few years earlier.) Similarly, breaks in work were measured by the minutes the electricity supply diminished. To form average hours of a group of workers, he took account of workers absent from the factory and he subtracted "lost time" from scheduled hours to arrive at actual hours of work.

---

3 .See US Department of Labor, Bureau of Labor Statistics (1917a–d and 1919), Bulletins 221, 222, 223, 230, and 249.

The mean value of actual weekly hours of work is about 87% of the mean value of scheduled weekly hours. Changes in scheduled hours brought about similar changes in actual hours worked. That is, if H_{jt} denotes the actual average hours worked by group j workers in the week ending at t, and if $(H_{jt})^{SC}$ denotes the hours scheduled for the group j workers in week ending t, then using observations over all workers and all weeks, the least-squares regression of the logarithm of H_{jt} on the logarithm of $(H_{jt})^{SC}$ yields the following results:

$$ln(H_{jt}) = -0.064 + 0.981 ln(H_{jt})^{SC}$$
$$(0.086)\,(0.021)$$

where the figures in parentheses are robust standard errors and $R^2 = 0.942$. The hypothesis that a given proportionate change in scheduled hours maps into the same proportionate change in actual hours cannot be rejected.

Vernon's data consist of weekly observations on average hours worked and index numbers of average output for each of four groups of workers: there are 56 weekly observations on 100 women turning fuze bodies, 27 weekly observations on 40 women milling a screw thread, 32 weekly observations on 56 men sizing fuze bodies, and 9 weekly observations on 15 youths boring top caps. In all, there are 124 weekly observations, the equivalent of over two years of weekly observations.[4] These data span the period from autumn 1915 to the end of 1916, during which "there were no changes whatever in the conditions of production of the articles, such as the character of the machinery and its speed, and in the nature and quality of the

---

4. There are two more weekly observations than those used in earlier work (Pencavel 2015b). The earlier research overlooked two weeks that are included in the present analysis. Both of the added observations are for the week ending September 2, 1916, one for the 40 women milling screw threads and the other for the 56 men sizing fuze bodies. The presence or absence of these two observations does not change inferences drawn.

articles produced" (Vernon 1921, 38). Between late 1915 and the end of 1916, there were changes in the length of the working week prompted by growing doubts about the wisdom of long hours: by the second half of 1916, work on Sunday had been eliminated for some workers and occasional holidays were introduced.[5]

Some of these workers were observed from mid-November 1915 to the end of December 1916. This raises the question of whether the skills of the workers improved over the weeks by a learning-by-doing process. Vernon was aware of this possibility, and he avoided using observations on recently hired workers whose inexperience might affect their productivity. He wrote of the workers,

> Since most of their work was of a comparatively simple repetition character, they soon attained a steady output. . . . [M]y method was to study the output of a group of experienced women for several months, and at the end of that time to replenish the reduced numbers [through turnover] by the addition of comparatively fresh workers. . . . It was found that fresh workers (provided they had one month's experience) achieved as great an output as experienced workers of five months' service so the introduction of the fresh blood did not appreciably alter output. (39)

Hours of Work and Output

Descriptive statistics on all the worker weeks and on each group of workers are provided in tables 4.1, 4.2 and 4.3.[6] The range of weekly hours worked is unusually wide: from 24 hours (in the week the

---

5. For instance, in the week ending 30 September 1916, the government called for a four-day general holiday.

6. Though Vernon (1917) sketches one figure relating hourly output to weekly hours of work of these workers, he presents no equations. To the best of my knowledge, the first regression equations fitted to these data are those in Pencavel (2015b).

Estimates of Production Functions

TABLE 4.1 Descriptive Statistics on Vernon's Weekly Observations of Hours and Output (All 124 Weekly Observations)

	Hours		Output
	Worked	Scheduled	
μ	50.4	58.0	6,275.5
min	24	29.5	2,812.4
Q_L	47.1	54.5	5,853.5
M	50.5	58.5	6,481.1
Q_U	55.5	66.5	6,990
max	72.5	78.5	8,735.2
σ	9.28	10.3	1,088.4
cv	0.184	0.178	0.173
$\Delta Q/M$	0.166	0.205	0.175

Note: The mean value of the dichotomous variable S_{jt} is 0.39.
The arithmetic mean is μ. The lowest value is *min* and the highest value is *max*. Q_L is the lower quartile (or 25th percentile) and Q_U is the upper quartile (or 75th percentile). M is the median.
The standard deviation is given by σ and the coefficient of variation by *cv*.

$\Delta Q/M$ is $(Q_U - Q_L)/M$.

government called for a four-day holiday) to 72.5 hours with a central tendency around 50 hours. Visual impressions of the relation between hours and output are provided by the scatter diagrams in figure 3 on the women workers and figure 4 on the men and youths.

After examining the output–hours association within each of the four groups of workers, the observations were pooled and, in estimating the regression equations, each group was allowed to have its own intercept (or fixed effect). To the 124 weekly

TABLE 4.2 Descriptive Statistics of Hours and Output of 56 Weekly Observations on 100 Women Turning Fuze Bodies and 27 Weekly Observations on 40 Women Milling a Screw Thread in Vernon's Data

	Women Turning Fuze Bodies			Women Milling a Screw Thread		
	Hours			Hours		
	Worked	Scheduled	Output	Worked	Scheduled	Output
μ	51.4	59.8	6,455.2	49.3	56.6	5,899.7
Min	24	29.5	2,812.4	26.4	29.5	3,141.6
Q_L	47.1	54	6,061.5	45	54.5	5,445
M	51.1	58.5	6,596.6	49.5	58.5	6,183.1
Q_U	56.1	66.5	7,162.9	55.4	66.5	6,490
max	69.2	77.3	8,735.2	64.9	71.8	7,062.5
σ	9.64	10.6	1,151.5	9.13	9.96	1,002.5
cv	0.188	0.177	0.178	0.185	0.176	0.170
$\Delta Q/M$	0.176	0.214	0.167	0.210	0.205	0.169

Note: The mean value of the dichotomous variable S_{jt} is 0.52 for the women turning fuze bodies and 0.37 for women milling a screw thread. The meaning of the symbols in the first column is provided beneath table 4.1.

observations collected by Vernon, the following equations are fitted by conventional least-squares:

(CD-1)* $\ln X_{jt} = \beta_0 + \beta_1 \ln H_{jt} + \omega_j + \varepsilon_{jt}$
(CD-2)* $\ln X_{jt} = \beta_0 + \beta_2 \ln H_{jt} + \beta_3 (\ln H_{jt})^2 + \omega_j + \varepsilon_{jt}$
(GZ-1)* $\ln X_{jt} = \alpha_0 + \alpha_1 H_{jt} + \omega_j + \varepsilon_{jt}$
(GZ-2)* $\ln X_{jt} = \alpha_0 + \alpha_1 H_{jt} + \alpha_2 (H_{jt})^2 + \omega_j + \varepsilon_{jt}$

where ω_j represents a fixed effect for each group of workers and ε_{jt} is a well-behaved stochastic disturbance term. Equations (CD-1)*, (CD-2)*, (GZ-1)*, and (GZ-2)* are the same as equations (CD-1), (CD-2), (GZ-1), and (GZ-2) in the previous chapter except that

TABLE 4.3 Descriptive Statistics of Hours and Output on 32 Weekly Observations on 56 Men Sizing Fuze Bodies and 9 Weekly Observations on 15 Youths Boring Top Caps in Vernon's Data

	Men Sizing Fuze Bodies			Youths Boring Top Caps		
	Hours			Hours		
	Worked	Scheduled	Output	Worked	Scheduled	Output
μ	47.9	54.5	6,174.1	56.8	62.7	6,644.9
min	27.1	29.5	3,614	47.4	51.1	5,877.6
Q_L	46.8	53.5	5,695.5	52.8	58.2	6,136.2
M	50.4	57.3	6,383.7	54.7	61.5	6,811.2
Q_U	51.6	58.5	6,877.8	56.2	63.4	7,193.6
max	59.1	66.7	7,738	72.5	78.5	7,324.6
σ	8.21	9.72	1,089.6	8.52	8.96	575.9
cv	0.171	0.178	0.176	0.150	0.143	0.087
$\Delta Q/M$	0.095	0.087	0.185	0.062	0.085	0.155

Note: The mean value of the dichotomous variable S_{jt} is 0.19 for the men sizing fuze bodies and 0.36 for the youths boring top caps. The meaning of the symbols in the first column is provided beneath table 4.1.

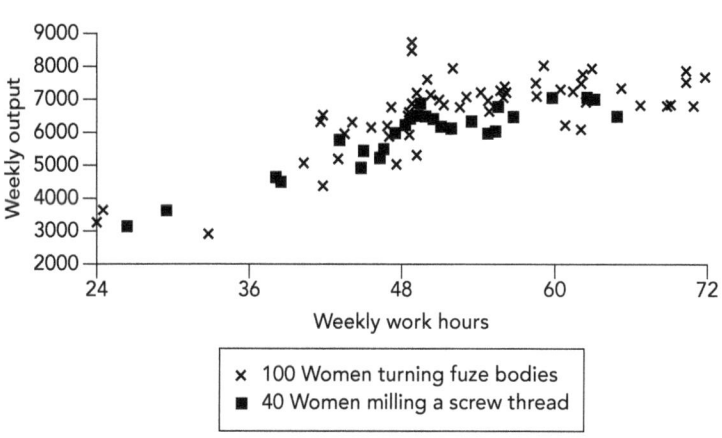

FIGURE 3 Weekly Output and Weekly Hours: 56 Observations on 100 Women Turning Fuze Bodies and 27 Observations on 40 Women Milling a Screw Thread on Fuze Bodies.

DIMINISHING RETURNS AT WORK

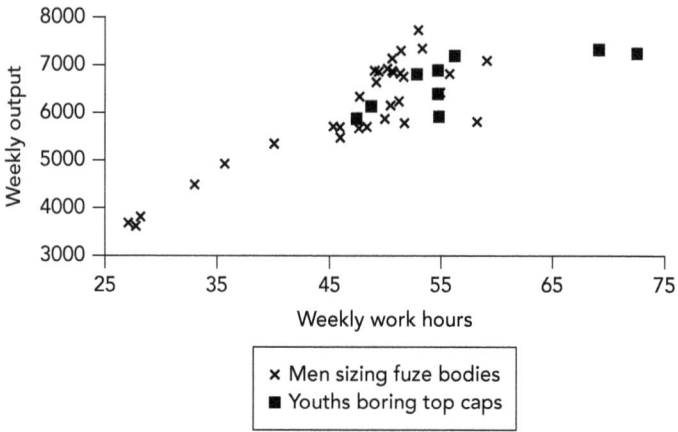

FIGURE 4 Weekly Output and Weekly Hours: 32 Observations on 56 Men Sizing Fuze Bodies and 9 Observations on 15 Youths Boring Top Caps.

those immediately above do not include the term $\eta ln Z_{jt}$. This is because, according to Vernon, over the period studied, there were no changes in production methods and in other inputs, such as the physical capital.

Equation (CD-1)* adopts Cobb and Douglas's specification where the elasticity of output with respect to changes in hours, $\mathscr{E}(X, H)$, is a constant, β_1. If β_1 lies between zero and unity, effort per hour declines as hours increase and the output–hours relation follows the law of diminishing returns. If β_1 is unity, effort per hour is independent of hours and hours are not subject to diminishing returns. Equation (CD-2)* allows the elasticity of output with respect to hours to vary with hours and a point estimate of this elasticity, $\mathscr{E}(X, H)$, is $\beta_2 + 2\beta_3 ln(H_{jt})$.

Equations (GZ-1)* and (GZ-2)* are inspired by Gompertz' semi-logarithmic specification. If effort per hour is independent of hours worked, α_2 will be zero and equation (GZ-1)* will describe the observations as well as equation (GZ-2)*. If α_1 is positive and α_2 is negative, hours follows the law of diminishing returns.

The least-squares parameter estimates of these equations are reported in table 4.4, where the estimate of β_1 in equation (CD-1)* is (by conventional criteria) significantly less than unity: a 10% cut in hours of work reduces output by 8%. This fails to confirm the hypothesis of unit returns to hours. Because the estimated value of β_1 is less than 1, output rises continuously with hours worked, but the output increase falls as hours lengthen; the law of diminishing returns to hours is satisfied. Because β_1

TABLE 4.4 Least-Squares Estimates of Equations (CD-1)*, (CD-2)*, (GZ-1)*, and (GZ-2)* Fitted to Vernon's 124 Weekly Observations

Right-hand-side variable ⇓	(CD-1)*	(CD-2)*	(GZ-1)*	(GZ-2)*
$\ln(H_{jt})$	0.814 (0.058)	5.437 (0.952)		
$[\ln(H_{jt})]^2$		−0.616 (0.125)		
H_{jt}			0.017 (0.002)	0.069 (0.006)
$(H_{jt})^2$				−0.00054 (0.00001)
goodness of fit statistics				
R^2	0.706	0.753	0.623	0.771
See	0.111	0.102	0.126	0.098

Note: The left-hand-side variable for each of the equations above is the natural logarithm of output. Estimated standard errors in parentheses are heteroskedastic-robust. The standard error of estimate of the fitted equation is see.

(CD-1)* $\ln X_{jt} = \beta_0 + \beta_1 \ln H_{jt} + \omega_j + \varepsilon_{jt}$
(CD-2)* $\ln X_{jt} = \beta_0 + \beta_2 \ln H_{jt} + \beta_3 (\ln H_{jt})^2 + \omega_j + \varepsilon_{jt}$
(GZ-1)* $\ln X_{jt} = \alpha_0 + \alpha_1 H_{jt} + \omega_j + \varepsilon_{jt}$
(GZ-2)* $\ln X_{jt} = \alpha_0 + \alpha_1 H_{jt} + \alpha_2 (H_{jt})^2 + \omega_j + \varepsilon_{jt}$

is positive but less than unity, effort per hour declines as hours lengthen.

The estimates of equations (CD-1)* and (CD-2)* imply the output–hours relations graphed in figure 5 and the marginal products of hours implied by these equations are portrayed in figure 6. The implied marginal product of hours declines with hours for both estimated equations, although the decline is palpable for equation (CD-2)*. Indeed, the implied marginal product of hours for equation (CD-2)* becomes negative at 83 hours, an hours figure beyond that observed in these data.

The estimates of equation (CD-2)* represent a statistically significant improvement over those of equation (CD-1)* and, according to equation (CD-2)*'s estimates, the implied value of $\mathscr{E}(X, H)$, the elasticity of output with respect to hours, falls

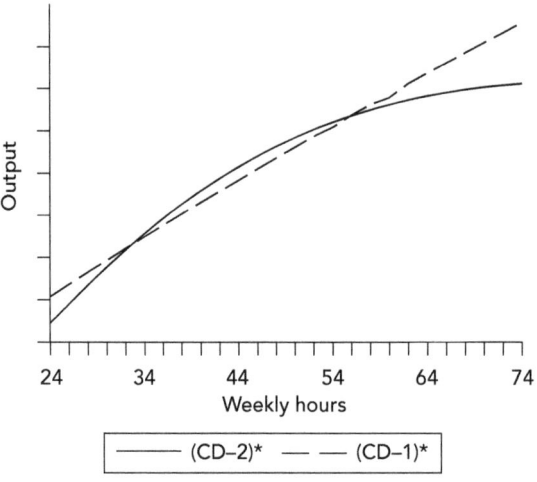

FIGURE 5 Output–Hours Relation Implied by the Estimates of Equations (CD-1)* and (CD-2)* Fitted to Vernon's 124 Weekly Observations

(CD-1)* $\quad \ln(X_{jt}) = \beta_0 + \beta_1 \ln(H_{jt}) + \omega_j + \varepsilon_{jt}$

(CD-2)* $\quad \ln(X_{jt}) = \beta_0 + \beta_2 \ln(H_{jt}) + \beta_3 [\ln(H_{jt})]^2 + \omega_j + \varepsilon_{jt}$

Estimates of Production Functions

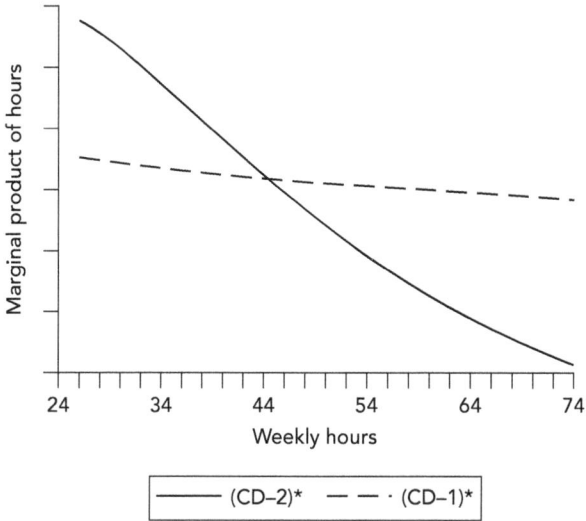

FIGURE 6 Marginal Product of Hours Implied by the Estimates of Equations (CD-1)* and (CD-2)* Fitted to Vernon's 124 Weekly Observations

(CD-1)* $\quad ln(X_{jt}) = \beta_0 + \beta_1 ln(H_{jt}) + \omega_j + \varepsilon_{jt}$

(CD-2)* $\quad ln(X_{jt}) = \beta_0 + \beta_2 ln(H_{jt}) + \beta_3 [ln(H_{jt})]^2 + \omega_j + \varepsilon_{jt}$

as hours increase, as shown in figure 7. The point estimate of $\mathcal{E}(X, H)$ is above unity at short hours and equals unity at 37 hours, after which it is below unity. According to the 95% confidence intervals drawn in figure 7, $\mathcal{E}(X, H)$ is not significantly different from unity between 32 and 40 hours.

Further evidence of the decline in $\mathcal{E}(X, H)$ as hours increase is provided by dividing the 124 observations into two regimes, a "short" hours regime and a "long" hours regime, and by estimating equation (CD-1)* to the observations within each regime separately. The consequences of this is shown by the estimates in table 4.5, where hours corresponding to three switching points are reported: 45 hours, 50 hours, and 55 hours. In all three cases, the estimated value of the coefficient β_1

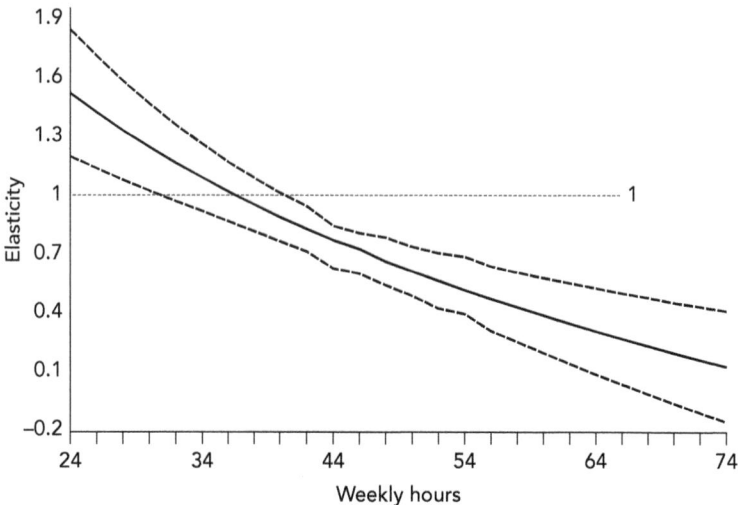

FIGURE 7 Estimated Values of $\mathscr{E}(X, H)$, the Effect of a Proportional Increase in Hours on $\ln(X)$, as Implied by Equation (CD-2)* Fitted to the 124 Observations Collected by Horace Vernon

(CD-2)* $\qquad \ln(X_{jt}) = \beta_0 + \beta_2 \ln(H_{jt}) + \beta_3 \left[\ln(H_{jt})\right]^2 + \omega_j + \varepsilon_{jt}$

The solid curve denotes $\mathscr{E}(X,H)$ and the dashed curves bounding $\mathscr{E}(X, H)$ represent 95% confidence intervals. The dotted horizontal line indicates a value of $\mathscr{E}(X, H)$ of unity.

in equation (CD-1)* is insignificantly different from unity in the shorter hours regime.

By contrast, for all three switching points, the estimated value of the coefficient β_1 is significantly less than unity in the longer hours regime. Indeed, the association between hours and output is much weaker in the longer hours regime These results suggest a nonlinear output–hours relation, with changes in hours having more meaningful effects on output at shorter hours than at longer hours.

Of the equations reported in table 4.4, as indicated by the standard error of estimate of the fitted equation (*see*), equation

Estimates of Production Functions

TABLE 4.5 Fitting Equation (CD-1)* to Subsets of Vernon's 124 Weekly Observations

Fitted to observations where H is ⇓	Estimated β_1	Estimated standard error of β_1	R^2	see	nobs
< 45	1.013	0.132	0.756	0.131	22
≥45	0.378	0.092	0.222	0.095	102
< 50	1.040	0.080	0.779	0.116	57
≥50	0.120	0.092	0.113	0.076	67
< 55	1.011	0.063	0.785	0.102	92
≥55	0.001	0.141	0.081	0.081	32

Note: The number of observations is *nobs*.

(CD-1)* $\qquad \ln X_{jt} = \beta_0 + \beta_1 \ln H_{jt} + \omega_j + \varepsilon_{jt}$

(GZ-2)* provides the best description of the observations. It implies the association between weekly hours and weekly output pictured in figure 8. The corresponding marginal product of hours (MP_H) and average product of hours (AP_H) implied by the estimates of equation (GZ-2)* are graphed in figure 9. Output per hour peaks at about 40 hours, and the estimated marginal product of hours exhibits the pattern drawn in introductory texts: it rises as hours increase from 24 to 34 hours, and then falls. Indeed, it becomes negative at 64 hours. Is a negative marginal product of hours at all plausible? In "Luddite" fashion,[7] were the workers destroying their output if they worked 64 or more hours?

---

[7]. The quotes here serve to distinguish the Luddites of common characterization—senseless opponents of the march toward mechanization—from the Luddites of historical fact. See Hobsbawm (1952).

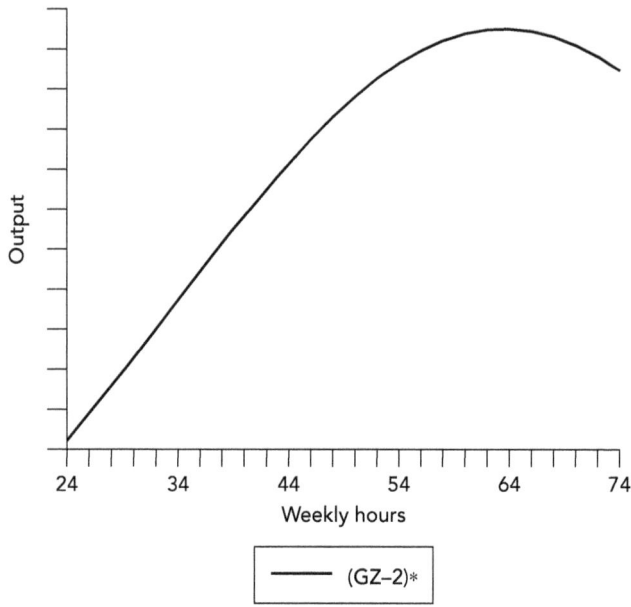

FIGURE 8 Output–Hours Relation Implied by the Estimates of Equation (GZ-2)* Fitted to the 124 Observations Collected by Horace Vernon

$$ln(X_{jt}) = \alpha_0 + \alpha_1 H_{jt} + \alpha_2 (H_{jt})^2 + \omega_j + \varepsilon_{jt}$$

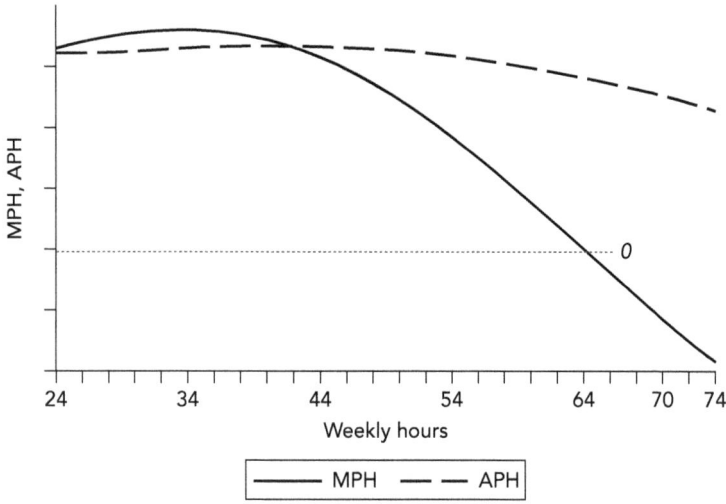

FIGURE 9 The Marginal Product of Hours (*MPH*) and Average Product of Hours (*APH*) Implied by the Estimates of Equation (GZ-2)* where $(MP_H)_{jt} = (\alpha_1 + 2.\alpha_2 H_{jt})X_{jt}$ Fitted to the 124 Observations Collected by Horace Vernon.

The dotted horizontal line denotes a zero value of the marginal product or average product of hours.

Estimates of Production Functions

No, this simply reflects the fact that weeks when high values of output were produced were often not weeks when hours were long, and if a single production function is fitted to these data, the production function will have a negative slope at long hours. Indeed, figure 10 shows the weeks with the highest levels of output for each group of workers and identifies the hours worked in these weeks. The weeks where output is highest are not the

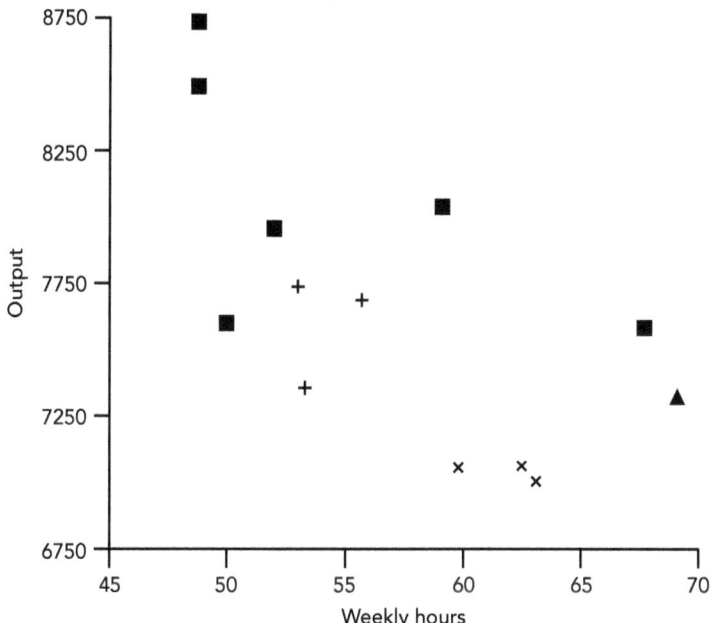

FIGURE 10 Weeks with Highest Weekly Output and their Corresponding Weekly Hours from Vernon's 124 Weekly Observations.

In this diagram, within each of the four groups, the weekly observations corresponding to the top ten percent of output have been recorded against their hours of work in that week. There are 6 observations for women turning fuze bodies denoted by ■, 3 observations on women milling screw threads denoted by ×, 3 observations on men sizing fuze bodies denoted by +, and 1 observation on youths boring top caps denoted by ▲.

weeks with the longest hours. Indeed, among this selected set of observations, longer hours are not associated with greater output.

Why would any employer knowingly schedule hours at which their marginal product is negative? The answer is that an employer may well not know or recognize that the marginal product is negative. The HMWC implied this shortcoming in employers' understanding when the committee reported they knew of no employer who, having once scheduled shorter hours, had subsequently rescinded them and returned to longer hours of work—the implication being that, before hours were reduced, employers were behaving suboptimally. The committee certainly believed that a reduction in hours would not reduce output (tacitly accepting the notion that workers' hours were at levels where the marginal product of hours was nonpositive). Thus, in its Final Report (HMWC 1919, p. 38, para. 156), the committee concluded "substantial reduction in hours can be effected without any reduction in output."

However, these negative marginal products are inferences from point estimates; they come with standard errors and they reflect, in part, a particular functional form. A conservative judgment would be that output is relatively unresponsive to increases in hours beyond 55, implying that a move from 65 hours to 55 hours would have second-order effects on the output of these workers.

The estimates of equation (GZ-2*) provide values for α_1 and α_2. When inserted into equation (6), they imply that effort per hour declines as hours increase and the effective hours of work follow a positive quadratic path in nominal hours.

Days of Work and Output

In the nineteenth century, many employers assumed that, if workers spend more time at work, more output would be forthcoming. This way of thinking induced manufacturers at the beginning of the First World War to return to the seven-day working week to satisfy the military's demands for armaments. Vernon's observations identify weeks when Sunday was a working day, and assuming that Sunday work connotes a seven-day working week, the effect of Sunday work on output and the effect of Sunday work on the output–hours relation can be investigated with these observations.[8]

Define a dichotomous variable S_{jt} that takes the value of unity when group j workers in week t worked on Sunday, and of zero when group j workers do not work on Sunday in week t. Unsurprisingly, hours of work are longer when seven days are worked: when $S_{jt} = 1$, average weekly hours are 10 hours longer than when $S_{jt} = 0$. Does a working week of seven days have consequences for output independent of hours or does the effect of seven days of work operate through hours only?

To answer this, consider the following regression equations that include the dichotomous variable S_{jt}:[9]

(9) $\quad \ln X_{jt} = \gamma_0 + \gamma_1 S_{jt} + \omega_j + \varepsilon_{jt}$

(10) $\quad \ln X_{jt} = \gamma_2 + \gamma_3 S_{jt} + \gamma_4 \ln H_{jt} + \omega_j + \varepsilon_{jt}$

(11) $\quad \ln X_{jt} = \gamma_5 + \gamma_6 S_{jt} + \gamma_7 \ln H_{jt} + \gamma_8 (\ln H_{jt})^2 + \omega_j + \varepsilon_{jt}$

(12) $\quad \ln X_{jt} = \delta_0 + \delta_1 S_{jt} + \delta_2 H_{jt} + \omega_j + \varepsilon_{jt}$

(13) $\quad \ln X_{jt} = \delta_3 + \delta_4 S_{jt} + \delta_5 H_{jt} + \delta_6 (H_{jt})^2 + \omega_j + \varepsilon_{jt}$

---

8. Sunday work may not imply a seven-day working week around holidays when Sunday might be a working day in lieu of a different day of rest. However, this will be an infrequent event. The Health of Munition Workers Committee interpreted Sunday work as a seven-day working week.

9. S_{jt} is a dichotomous variable for all but four of the weekly observations. For these four, S_{jt} lies between zero and unity. All four observations relate to the 15 youths boring top caps and S_{jt} is the fraction of these youths working on Sunday in week ending t.

Again, ω_j is a fixed effect for each group of workers and ε_{jt} is a stochastic residual. Equation (9) is designed to document the association between S_{jt} and output. Equations (10) through (13) mimic equations (CD-1)*, (CD-2)*, (GZ-1)*, and (GZ-2)* with the addition of the Sunday dichotomous variable S_{jt}.[10]

The least-squares estimates of these equations are given in table 4.6. The estimates of equation (9) show that, when hours are excluded, output and days of work are uncorrelated. Whenever hours of work are entered on the right-hand side of the estimating equation in addition to S_{jt}, the impact of Sunday work is unambiguously negative: holding working hours constant, Sunday work is associated with between 13.5 and 17.4% lower weekly output. The damaging effect of a seven-day workwork was identified by Thomas Loveday (1917, p. 44, para. 2b), another investigator for the Health of Munition Workers Committee, who wrote, "The effects of Sunday labour are, as has now been recognized, still worse than those of overtime hours in the evening or on Saturday afternoon."

The estimates attached to the hours variables after controlling for days of work are similar to those not controlling for S_{jt} although the presence of S_{jt} attenuates some of the effects on output of long hours.[11] The estimates attached to δ_5 and δ_6 in equation (13) suggest the same positive concave relationship between output and hours as that shown in figure 8 for the estimates of

---

10. The addition of S_{jt} may be accommodated in the models in chapter 3 as follows: For the specification inspired by Cobb and Douglas, suppose in equation (3) $b_{jt} = exp(c_1 \cdot S_{jt})$. Then equations (CD-1) and (CD-2) have the term $c_1 \cdot S_{jt}$ added to the right-hand side. For the specification inspired by Gompertz, we may write equation (2) as $L = E.H + c_2 S$ and the term $c_2 S_{jt}$ will be added to the right-hand side of equations (GZ-1) and (GZ-2).

11. If effort takes the form of $E_{jt} = a_1 + a_2 H_{jt} + a_3 S_{jt}$, then an interaction term $S_{jt} \cdot H_{jt}$ will appear in the expression for L_{jt} and in Gompertz's production function. This specification was also estimated with consequences very similar to those when S_{jt} appears additively as in equation (13).

Estimates of Production Functions

TABLE 4.6 Least-Squares Estimates of the Effect on Output of a Seven Day Working Week Fitted to Vernon's 124 Weekly Observations

Right-hand-side variable	Equation (9)	Equation (10)	Equation (11)	Equation (12)	Equation (13)
	Robust standard errors in parentheses below coefficient estimates				
S_{jt}	0.0398	−0.168	−0.149	−0.191	−0.145
	(0.036)	(0.022)	(0.025)	(0.026)	(0.023)
$\ln H_{jt}$		0.999	3.018		
		(0.051)	(0.870)		
$(\ln H_{jt})^2$			−0.272		
			(0.118)		
H_{jt}				0.022	0.062
				(0.002)	(0.005)
$(H_{jt})^2$					−0.00042
					(0.00001)
Goodness of fit statistics					
R^2	0.047	0.822	0.829	0.762	0.843
see	0.200	0.087	0.085	0.100	0.082

(9) $\quad \ln X_{jt} = \gamma_0 + \gamma_1 S_{jt} + \omega_j + \varepsilon_{jt}$

(10) $\quad \ln X_{jt} = \gamma_2 + \gamma_3 S_{jt} + \gamma_4 \ln H_{jt} + \omega_j + \varepsilon_{jt}$

(11) $\quad \ln X_{jt} = \gamma_5 + \gamma_6 S_{jt} + \gamma_7 \ln H_{jt} + \gamma_8 (\ln H_{jt})^2 + \omega_j + \varepsilon_{jt}$

(12) $\quad \ln X_{jt} = \delta_0 + \delta_1 S_{jt} + \delta_2 H_{jt} + \omega_j + \varepsilon_{jt}$

(13) $\quad \ln X_{jt} = \delta_3 + \delta_4 S_{jt} + \delta_5 H_{jt} + \delta_6 (H_{jt})^2 + \omega_j + \varepsilon_{jt}$

equation (GZ-2)*. According to the estimates of equation (13), output reaches a maximum when hours reach 74 (just beyond the highest value of hours in these data), after which the marginal product of hours is negative.

The harm implied for output from a seven-day working week may be illustrated by comparing two combinations of daily hours and days of work: one involves a seven-day working week with 10 hours worked on each day; the other entails work on six days with 9 hours worked on each of the six days. The estimates of equation (13) imply that these two combinations of daily hours and weekly days yield about the same output, although the shorter working week would have been effected at lower cost to the employer and with more opportunity for leisure and repair for the workers.

Recovery from Work

What is it about seven working days in a week that is so noxious? The HMWC determined that the damage wrought by Sunday labor arose because it prevented workers from securing their weekly rest and inhibited recovery from fatigue.[12] This notion that, to be effective in the workplace workers require time to recover from work, has been taken up in recent years by occupational health psychologists who have documented the deleterious consequences of inadequate recuperation.[13]

The estimates of the regression equations just reported show that output in week t is harmed by Sunday work in week t. If the committee's explanation is correct, and if Sunday work is followed by work the following day (Monday), then not merely is output in week t damaged by Sunday work in week t but also Sunday work in the previous week $t-1$ will harm output in week t because the

---

12. The precise statement is "The evidence is conclusive that Sunday labour by depriving the worker of his weekly rest offers him no sufficient opportunity for recovering from fatigue, and is not productive of greater output except for quite short and isolated periods. . . . [S]even days' labour only produces six days' output and . . . reductions in Sunday work have not in fact involved any appreciable loss of output" (HMWC 1919, p. 44, para. 194).

13. See, for instance, Fritz et al. (2010), Jansen et al. (2002), and Sonnentag and Zijlstra (2006).

workers in week t start the week's work without having recovered fully from the previous week of work. Many of Vernon's 124 weekly observations relate to contiguous weeks, so that information is available not merely on the current week's output and work but also on the previous week's output and work.

The hypothesis is that a long working week in week t causes workers to be fatigued in week t and also leaves them with inadequate time for repair before a new working week begins.[14] This harms output in week $t + 1$. To test this hypothesis, all those weeks that provide information on hours and days of work in the previous week are drawn from Vernon's 124 weeks. There are 108 such weeks, which represent about 87% of Vernon's 124 weekly observations. Descriptive statistics on the key variables on these 108 contiguous weekly observations are given in table 4.7. Comparing the entries in table 4.7 with those in table 4.1, we see that the distribution of the hours and output variables in this set of 108 observations is similar to that in the larger set of 124 observations.

Moreover, the inferences from fitting equations (CD-1)*, (CD-2)*, (GZ-2)*, (10), (11), and (13) to these 108 observations (as given in table 4.8) resemble those in tables 4.4 and 4.6 that were estimated using all 124 observations.

To examine the hypothesis that a long working week in week $t-1$ will have damaging consequences for output in week t, a variable is constructed that combines the effect of seven days of work and long hours in week $t-1$. Precisely, this variable, F_{jt-1}, is defined as the sum of S_{jt-1} and the ratio of actual hours in week $t-1$ to average hours over all observations.

---

14. Biddle and Hamermesh (1990) provide evidence that longer time at market work is associated with shorter time asleep.

TABLE 4.7 Descriptive Statistics of Hours and Output from the Set of 108 Weekly Observations Drawn from Vernon's 124 Weekly Observations Examining the Lagged Effects of a Long Working Week

	Hours worked	Output	F_{jt-1}
μ	49.4	6,245.4	1.33
min	24.0	2,812.4	0
Q_L	46.0	5,741.6	0.95
M	49.9	6,493.0	1.03
Q_U	54.5	6,990.0	2.06
max	69.2	8,735.2	2.38
σ	9.16	1,146.2	0.63
cv	0.185	0.184	0.474
$\Delta Q/M$	0.170	0.192	1.078

Note: The mean value of the dichotomous variable S_{jt} is 0.35.
F_{jt-1} is defined as $S_{jt-1} + (H_{jt-1}/50)$.
The meaning of the symbols in the first column is provided beneath table 4.1.

Average hours are approximately 50 (the arithmetic mean and the median), so

(14) $$F_{jt-1} = S_{jt-1} + \left[(H_{jt-1})/(50)\right].$$

To illustrate some values of F_{jt-1}, the women turning fuze bodies in the week ending 30 September 1916, worked only 24 hours, as the government asked for a four-day holiday and they did not work on Sunday. Hence, for the following week ending 7 October 1916, F_{jt-1} takes the value of 0.48, a relatively low value implying the workers started work in the following week (the week ending 7 October) having largely recovered from work the previous week. By contrast, in the week ending 5 December 1915, these same workers worked

TABLE 4.8 Least-Squares Estimates of Equations (CD-1)*, (CD-2)*, (GZ-2)*, (10), (11), and (13) Fitted to Vernon's 108 Observations with Information on Hours and Days of Work in Contiguous Weeks

Right-hand-side variable ⇓	Equation (CD-1)*	Equation (CD-2)*	Equation (GZ-2)*	Equation (10)	Equation (11)	Equation (13)
			Robust standard errors in parentheses below coefficient estimates			
S_{jt}				−0.164 (0.025)	−0.148 (0.028)	−0.139 (0.026)
$\ln H_{jt}$	0.861 (0.060)	5.166 (1.125)		1.012 (0.053)	2.770 (1.061)	
$(\ln H_{jt})^2$		−0.577 (0.150)			−0.238 (0.145)	
H_{jt}			0.0717 (0.006)			0.063 (0.005)
$(H_{jt})^2$			−0.0006 (0.0001)			−0.0004 (0.0001)
Goodness of fit statistics						
R^2	0.735	0.770	0.793	0.832	0.837	0.852
See	0.111	0.104	0.098	0.089	0.088	0.084

(CD-1)* $\ln X_{jt} = \beta_0 + \beta_1 \ln H_{jt} + \omega_j + \varepsilon_{jt}$
(CD-2)* $\ln X_{jt} = \beta_0 + \beta_2 \ln H_{jt} + \beta_3 (\ln H_{jt})^2 + \omega_j + \varepsilon_{jt}$
(GZ-2)* $\ln X_{jt} = \alpha_0 + \alpha_1 H_{jt} + \alpha_2 (H_{jt})^2 + \omega_j + \varepsilon_{jt}$
(10) $\ln X_{jt} = \gamma_2 + \gamma_3 S_{jt} + \gamma_4 \ln H_{jt} + \omega_j + \varepsilon_{jt}$
(11) $\ln X_{jt} = \gamma_5 + \gamma_6 S_{jt} + \gamma_7 \ln H_{jt} + \gamma_8 (\ln H_{jt})^2 + \omega_j + \varepsilon_{jt}$
(13) $\ln X_{jt} = \delta_3 + \delta_4 S_{jt} + \delta_5 H_{jt} + \delta_6 (H_{jt})^2 + \omega_j + \varepsilon_{jt}$

all seven days and they worked 68.4 hours. This implies that, for the week ending 12 December 1915, the value of F_{jt-1} is 2.368, almost the maximum value of F_{jt-1}, implying little opportunity for recovery from work before starting work the following week. So higher values of F_{jt-1} indicate greater fatigue and inadequate recovery from the previous week's work, and this is hypothesized to damage worker performance the following week, week t.[15] A high value of F_{jt-1} suggests both a fatiguing working week in $t-1$ and little time for recovery before work starts in week t.

On four occasions, the entire week in $t-1$ was a holiday, so S_{jt-1} and H_{jt-1} are zero and thus F_{jt-1} is also zero. This value of F_{jt-1} is an outlier. To ensure the results that follow are not unduly influenced by these outlier values, all the equations reported were fitted also to the 104 weeks for which the values of all observations on F_{jt-1} are positive. The inferences from equations fitted to the 104 observations that omit weeks when F_{jt-1} is zero are very similar to those fitted to the 108 weekly observations that include zero values of F_{jt-1}.

Note that there is positive serial correlation in the length of work in one week and the length of work in the following week. For instance, of the 30 weeks with hours of work greater than 54 hours, two-thirds were preceded by a working week of 54 hours or more. In more than 90% of weeks, $S_{jt} = S_{jt-1}$ with the breakdown between $S = 0$ and $S = 1$ given in table 4.9. The most frequent pattern (65 weeks) is when both S_{jt} and S_{jt-1} are zero: most of the contiguous weeks did not involve work on Sunday. Seven-day working weeks are more frequent before the autumn of 1916.

Now, augment equations (10), (11), and (13) with F_{jt-1}:

(10 $_F$) $\ln X_{jt} = \gamma_9 + \gamma_{10} F_{jt-1} + \gamma_{11} S_{jt} + \gamma_{12} \ln H_{jt} + \omega_j + \varepsilon_{jt}$

(11 $_F$) $\ln X_{jt} = \gamma_{13} + \gamma_{14} F_{jt-1} + \gamma_{15} S_{jt} + \gamma_{16} \ln H_{jt}$
$\qquad + \gamma_{17} (\ln H_{jt})^2 + \omega_j + \varepsilon_{jt}$

(13 $_F$) $\ln X_{jt} = \delta_7 + \delta_8 F_{jt-1} + \delta_9 S_{jt} + \delta_{10} H_{jt} + \delta_{11} (H_{jt})^2 + \omega_j + \varepsilon_{jt}$

---

15. See Pencavel (2016b) for an analysis that unbundles F_{jt-1} and examines the separate effects of S_{jt-1} and H_{jt-1} on output in week t.

TABLE 4.9 The Association between S_{jt} and S_{jt-1} among Vernon's 108 Contiguous Weeks

	$S_{jt-1}=1$	$S_{jt-1}=0$	Row total
	Number of weeks		
$S_{jt}=1$	33	4	37
$S_{jt}=0$	6	65	71
Column total	39	69	108

If the committee's judgment is correct that adequate work performance in week t requires recovery from the previous week's work, holding constant the current week's hours and days of work, higher values of F_{jt-1} (greater fatigue from the previous week's work) will damage output in week t and the coefficients attached to F_{jt-1} will be negative. The least-squares estimates of the parameters of equations (10_F), (11_F), and (13_F) with robust standard errors attached to the estimated coefficients are given in table 4.10.

These estimates imply that, holding hours and days of work in week t constant, a long working week in the previous week (that is, in week $t-1$) reduces output in week t. When F_{jt-1} assumes its mean value of 1.33, output in week t is lower by between 4 and 5.2 %. When F_{jt-1} takes on its maximum value of 2.38, output in week t is depressed by 7 to 10%. These results imply that there are intertemporal effects in the output–hours relation. This should not occasion surprise: just as machines depreciate with use and require care and maintenance to remain productive, so women and men tire at work and require time outside of work to restore their minds and bodies for further work. Without that restoration, their effective labor depreciates.

This was well expressed by the HMWC (1919, p. 40, para. 162) when it noted that not only would shorter hours of work result in

DIMINISHING RETURNS AT WORK

TABLE 4.10 Least Squares Estimates of Equations (10_F), (11_F), and (13_F) to Vernon's Set of 108 Observations

Right-hand-side variable	Equation (10_F)	Equation (11_F)	Equation (13_F)
	Robust standard errors in parentheses next to coefficient estimates		
F_{jt-1}	−0.041 (0.015)	−0.037 (0.015)	−0.030 (0.015)
S_{jt}	−0.125 (0.025)	−0.115 (0.028)	−0.113 (0.027)
$\ln H_{jt}$	1.016 (0.053)	2.504 (1.026)	
$(\ln H_{jt})^2$		−0.201 (0.141)	
H_{jt}			0.061 (0.005)
$(H_{jt})^2$			−0.0004 (0.0001)
Goodness of fit statistics			
R^2	0.839	0.842	0.856
See	0.088	0.087	0.083

(10_F) $\quad \ln X_{jt} = \gamma_9 + \gamma_{10} F_{jt-1} + \gamma_{11} S_{jt} + \gamma_{12} \ln H_{jt} + \omega_j + \varepsilon_{jt}$

(11_F) $\quad \ln X_{jt} = \gamma_{13} + \gamma_{14} F_{jt-1} + \gamma_{15} S_{jt} + \gamma_{13} \ln H_{jt} + \gamma_{14} (\ln H_{jt})^2 + \omega_j + \varepsilon_{jt}$

(13_F) $\quad \ln X_{jt} = \delta_7 + \delta_8 F_{jt-1} + \delta_9 S_{jt} + \delta_{10} H_{jt} + \delta_{11} (H_{jt})^2 + \omega_j + \varepsilon_{jt}$

lower costs for the employer but also they would provide the worker with "a longer period of recovery, for the enjoyment of sleep and rest, and for the necessary opportunity for recreation, exercise, and the discharge of the ordinary duties of citizenship and domestic life."

Some Conclusions About Hours, Days, and Output of Munition Workers in the First World War

This analysis of the observations collected by Horace Vernon on the British munition workers during the Great War leads to the conclusion that hours of work enters the production function in a decidedly nonlinear manner. At "short" hours, output is relatively sensitive to changes in hours worked and, over some hours (approximately 32 to 40 weekly hours), the hypothesis of *unit returns to hours* cannot be rejected. It can be rejected for "long" hours, where output is relatively insensitive to changes in hours. Indeed, the elasticity of output with respect to hours, $\mathscr{E}(X, H)$—that is, the relationship between proportional changes in hours and proportional changes in output—is not a constant. Neither the hypothesis of *unit returns to hours* nor the hypothesis of *constant returns to hours* is supported by these observations.

If the equations estimated with Vernon's observations applied to workers in the nineteenth century, the weekly hours of many of these workers were at levels where the effect of a reduction in hours would have had small, perhaps negligible, effects on output. In short, the claim by many economists of the time that a mandatory reduction in hours would result in severe negative consequences for output is not supported by the production functions fitted to Vernon's observations.

Observations Collected by the Industrial Health Research Board During the Second World War

The concern in Britain for the health and productivity of workers in the munition factories during the First World War reappeared in the Second World War. Once again, at the outbreak of war, Britain's military preparedness—including its stocks of ordnance—fell short of the enemy's. To address the deficiency, hours of work in British munition factories were extended. As before, young men left for the military and their places in the factories were taken by new entrants to the labor market, many of them women.

In a manner similar to Horace Vernon's investigations in the Great War, the Industrial Health Research Board (IHRB), whose lineage can be traced to the HMWC, undertook an investigation of workers' hours in munition factories and their output. The factories selected were those where "no significant changes, apart from hours, in the type or conditions of the work" had been made since the beginning of the war (IHRB 1942, 11). The observations were restricted to "a limited number of experienced workers who were regularly employed on the same type of work" (11).

Weekly hours and weekly output are averaged over the weeks that constitute each month from April 1940 to April 1941. Variations in hours of work reflect occasional holidays and the interruptions caused by bombing and air-raid warnings. The period studied includes major Luftwaffe raids on London, Birmingham, Coventry, West Bromwich, Southampton, Bristol, Sheffield, Liverpool, Manchester, Clydebank, Plymouth, and Belfast. There are six factories in the IHRB data. The workers are men in five factories and women in the sixth. There are 12 monthly observations per factory for five factories and 11 monthly observations in the sixth factory; there are 71 observations in all. The identity and location of these factories are not provided.

Figure 11 presents a scatter diagram of the observations on output versus hours and table 4.11 provides descriptive statistics on hours and output. Although the central tendency of hours is similar to that in Vernon's data, the range of hours in these Second World War observations is narrower than those from the First World War; this may be due, in part, to these data being averaged over the weeks that constitute a month.

Estimates of Production Functions

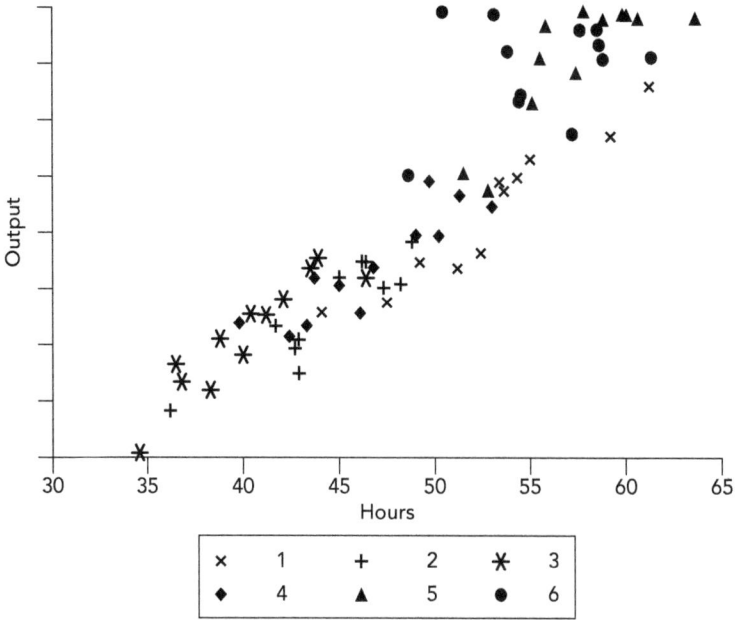

FIGURE 11 Monthly Observations Compiled by the Industrial Health Research Board on Six Groups of Workers: Their Average Weekly Output mapped against their Average Weekly Hours of Work.

Group 1 consists of 115 women workers, group 2 consists of 74 men, group 3 consists of 190 men, group 4 consists of another 190 men, group 5 consists of 200 men, group 6 consists of another 200 men.

Defining X_{jt} as an index number of plant j's weekly output averaged over the weeks in month t and H_{jt} as weekly hours worked in plant j averaged over the weeks in month t, the following equations were fitted by conventional least-squares to the 71 observations:

(CD-1)* $\qquad ln X_{jt} = \beta_0 + \beta_1 ln H_{jt} + \omega_j + \varepsilon_{jt}$

(CD-2)* $\qquad ln X_{jt} = \beta_0 + \beta_2 ln H_{jt} + \beta_3 (ln H_{jt})^2 + \omega_j + \varepsilon_{jt}$

(GZ-1)* $\quad\quad \ln X_{jt} = \alpha_0 + \alpha_1 H_{jt} + \omega_j + \varepsilon_{jt}$

(GZ-2)* $\quad\quad \ln X_{jt} = \alpha_0 + \alpha_1 H_{jt} + \alpha_2 (H_{jt})^2 + \omega_j + \varepsilon_{jt}$

Here, ω_j represents a fixed effect for each plant and ε_{jt} is an assumed well-behaved stochastic disturbance term. The least-squares estimates of these four equations are reported in table 4.12. All four specifications remove a relatively large fraction of the variations in the logarithm of output and, on this criterion, there is little to choose among them.

The estimate of β_1 in equation (CD-1)* is close to unity and supports the hypothesis of constant returns to hours: an x% reduction in hours induces approximately an x% reduction in output. However, the standard error of estimate of the fitted equation is least for the two equations that allow for a quadratic term in hours. The output–hours relation implied by the estimates of equations

TABLE 4.11 Descriptive Statistics on Weekly Hours and Weekly Output of the 71 Observations Reported by the IHRB

	μ	min	Q_L	M	Q_U	max	σ	cv	$\Delta Q/M$
X_{jt}	51.36	32.4	43.2	48.8	60.9	66.8	9.75	0.190	0.363
H_{jt}	49.4	34.6	43.3	49.2	55.1	63.6	7.31	0.148	0.240

Note: X_{jt} is an index number of plant j = s weekly output averaged over the weeks in month t.
H_{jt} is weekly hours worked in plant j averaged over the weeks in month t.
The arithmetic mean is denoted by μ, the standard deviation by σ, the lower quartile by Q_L, the median by M, and the upper quartile by Q_U. The coefficient of variation is cv. The minimum and maximum values of each variable are given under *min* and *max*, respectively.
$\Delta Q/M$ is $(Q_U - Q_L)/M$.

Estimates of Production Functions

TABLE 4.12 Least-Squares Estimates of the Relationship between Working Hours and Output Using Observations Collected by the IHRB

Right-hand-side variable ⇓	Estimates (standard errors) of equation			
	........(CD-1)*	........(CD-2)*	........(GZ-1)*	........(GZ-1)*
H_{jt}			0.020 (0.002)	0.061 (0.014)
$(H_{jt})^2$				−0.0004 (0.0001)
$\ln H_{jt}$	0.987 (0.081)	4.492 (2.621)		
$(\ln H_{jt})^2$		−0.454 (0.341)		
Goodness of fit statistics				
R^2	0.934	0.936	0.926	0.936
See	0.0512	0.0507	0.0542	0.0507

Note: The left-hand-side variable for each of the equations above is the natural logarithm of output.

Estimated standard errors in parentheses are heteroskedastic-robust.

see is the standard error of estimate of the fitted equation.

(CD-1)* $\ln X_{jt} = \beta_0 + \beta_1 \ln H_{jt} + \omega_j + \varepsilon_{jt}$

(CD-2)* $\ln X_{jt} = \beta_0 + \beta_2 \ln H_{jt} + \beta_3 (\ln H_{jt})^2 + \omega_j + \varepsilon_{jt}$

(GZ-1)* $\ln X_{jt} = \alpha_0 + \alpha_1 H_{jt} + \omega_j + \varepsilon_{jt}$

(GZ-2)* $\ln X_{jt} = \alpha_0 + \alpha_1 H_{jt} + \alpha_2 (H_{jt})^2 + \omega_j + \varepsilon_{jt}$

(CD-2)* and (GZ-2)* are pictured in figure 12, with the implied marginal product of hours of each portrayed in figure 13. The implied output–hours relations of equations (CD-2)* and (GZ-2)* are similar, as are their implied marginal products of hours.

When examining the weekly observations collected by Vernon for the HMWC, it was found that, when fitted to

DIMINISHING RETURNS AT WORK

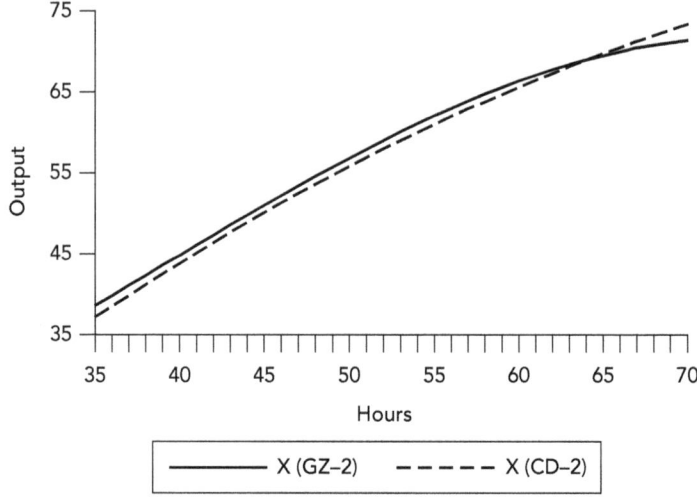

FIGURE 12 The Output–Hours Relation Implied by the Estimates of Equations (CD-2)* and (GZ-2)* Fitted to the Industrial Health Research Board's Observations

(CD-2)* $\ln X_{jt} = \beta_0 + \beta_2 \ln H_{jt} + \beta_3 (\ln H_{jt})^2 + \omega_j + \varepsilon_{jt}$

(GZ-2)* $\ln X_{jt} = \alpha_0 + \alpha_1 H_{jt} + \alpha_2 (H_{jt})^2 + \omega_j + \varepsilon_{jt}$

observations at all hours, the estimates of equation (CD-1)* masked different effects corresponding to "shorter" and "longer" hours. Is there evidence of this also in these observations from the IHRB in the Second World War? To answer this, the observations were divided into two regimes based on the value of weekly hours, and equation (CD-1)* was fitted to observations in each of the two regimes. The estimates that follow correspond to those observations where weekly hours are less than 54 and to observations where hours are equal to or greater than 54 hours.[16] The least-squares estimates of equation (CD-1)* fitted

---

16. Other definitions of the switching point in hours were considered with qualitatively similar implications to those reported where the threshold occurs at 54 hours.

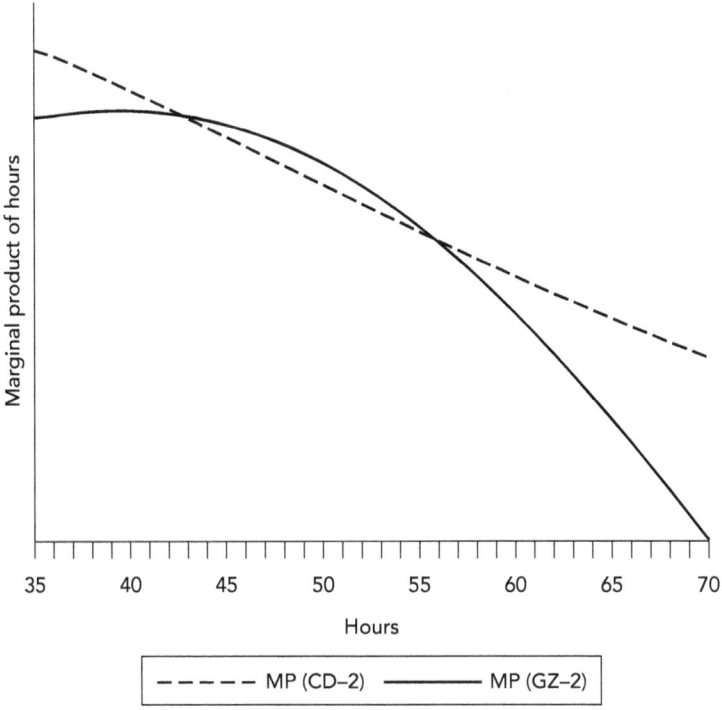

FIGURE 13 The Marginal Product of Hours as Implied by the Estimates of Equations (CD-2)* and (GZ-2)* Fitted to the Industrial Health Research Board's Observations

(CD-2)* $\quad \ln X_{jt} = \beta_0 + \beta_2 \ln H_{jt} + \beta_3 (\ln H_{jt})^2 + \omega_j + \varepsilon_{jt}$

(GZ-2)* $\quad \ln X_{jt} = \alpha_0 + \alpha_1 H_{jt} + \alpha_2 (H_{jt})^2 + \omega_j + \varepsilon_{jt}$

to the regimes of shorter hours and longer hours separately are reported in table 4.13.

The hypothesis of constant returns to hours cannot be rejected for the shorter hours regime, but this hypothesis can be rejected for the longer hours regime where the estimates of equation (CD-1)* imply that a 10% increase in hours worked raises output by 6.5%. The estimate of β_1 in equation (CD-1)* reported in

TABLE 4.13 Least-Squares Estimates (and Estimated Standard Errors in Parentheses) of Equation (CD-1)* Fitted to Industrial Health Research Board Observations in the Shorter Hours Regime and to Observations in the Longer Hours Regime Separately

Right-hand-side variable ⇓	Hours < 54	Hours ≥ 54
$\ln H_{jt}$	1.059	0.652
	(0.096)	(0.150)
R^2	0.910	0.781
see	0.0468	0.0330
nobs	49	22

Note: Estimated standard errors robust to heteroskedasticity are in parentheses beneath estimated coefficients. The number of observations is given by *nobs*.

table 4.12 (namely, 0.987) lies between those reported in table 4.13 (namely, 1.059 and 0.652) suggesting that, when estimated with all observations, the specification of equation (CD-1)* does not allow for different effects of hours on output when hours are long from those effects at short hours.

The point elasticity of output with respect to hours, $\mathcal{E}(X, H)$ (that is, the effect of a small proportionate increase in hours on proportional changes in output) is $\beta_2 + 2.\beta_3(\ln H_{jt})$ in equation (CD-2)*. Using the estimates of equation (CD-2)* in table 4.12, the implied values of $\mathcal{E}(X, H)$ are graphed as a function of H in figure 14. The point estimate of $\mathcal{E}(X, H)$ is greater than unity at hours less than 46 and not significantly different from unity between 42 and 52 hours. This range corresponds to slightly longer hours than was the case for Vernon's munitions workers in the First World War, but the presence of a nonlinearity is duplicated. As was the case

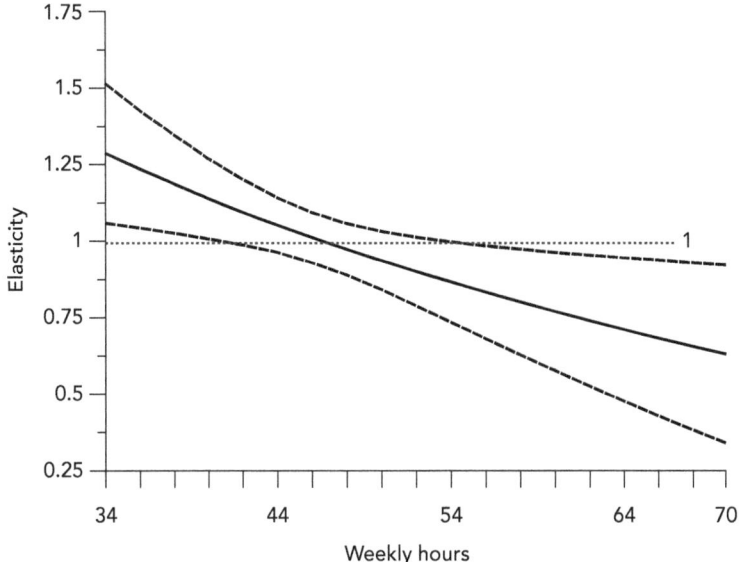

FIGURE 14 Estimated Values of $\mathscr{E}(X, H)$, the Effect of a Proportional Increase in Hours on $ln\,(X)$, as Implied by Equation (CD-2)* Fitted to the 71 Observations Collected by the Industrial Health Research Board

(CD-2) $\qquad ln(X_{jt}) = \beta_0 + \beta_2 ln H_{jt} + \beta_3 (ln H_{jt})^2 + \omega_j + \varepsilon_{jt}$

The dashed lines denote 95% confidence intervals around $\mathscr{E}(X, H)$. The dotted horizontal line indicates a value of $\mathscr{E}(X, H)$ of unity.

with Vernon's observations, the decline in $\mathscr{E}(X, H)$ as hours rise confirms the textbook figure of the ratio of the marginal product to average product illustrated in figure 2.

Vernon's observations in the First World War discussed earlier in the chapter were drawn from a single munitions factory. The IHRB's observations analyzed in this section describe six munitions factories in the Second World War. As both sets of observations relate to the production of munitions, are the output–hours relations fitted to Vernon's observations indistinguishable from those fitted to the IHRB's observations? For this, compare the estimates of (CD-2)* in table 4.4 with the estimates of (CD-2)* in table 4.12; a second comparison is of the estimates of (GZ-2)* in table 4.4 with the estimates of (GZ-2)* in table 4.12. Over the range of hours observed for both sets of observations (that is, from 35 to 64 hours), and for a given *form* of the production function (either Cobb and Douglas's augmented specification [(CD-2)] or Gompertz's augmented specification [(CD-2)]), the equations fitted to Vernon's data reveal more curvature than those estimated with the IHRB's data. The estimated production functions are similar but distinguishable.

There are two reasons for their being distinguishable: (1) the equations fitted to Vernon's observations are estimated over a wider range of hours, allowing the curve to absorb the effects of influential outliers; (2) although both data sets concern the production of munitions, those in the Second World War were substantially more sophisticated armaments as to constitute essentially different products involving a different production technology. In light of this, the surprising finding is not that they are distinguishable but that they are similar.

Observations Collected by Max Kossoris and Reinfried Kohler During the Second World War

In America, studies were undertaken during the Second World War by researchers in the Department of Labor into the association between hours and output; and shortly after the war, they were brought together in a bulletin titled "Hours of Work and Output" prepared by Max Kossoris and Reinfried Kohler (1948), subsequently referred to here as K&K.[17] They reported not only on working hours and output but also on hours and absenteeism and hours and work injuries.[18]

K&K's unit of observation is the plant or the department of a plant where the work or the workers form a distinct unit. The plants they selected for study were not random but, rather, had the following features: the work schedules were maintained long enough to reduce the probability of measuring merely transitory responses to changes in these schedules; the nature of the work did not change over the period studied; and output was measurable with records on hours, output, and other key variables. Their case studies cover 2,445 men and 1,060 women workers in 34 plants, in a diverse set of industries including metal products, electronic equipment, machinery, airplanes, shells, rubber products, apparel, and drugs.

With such a wide range of industries and, one assumes, of production technologies, it would be remarkable if a single type of function describing the conversion of inputs (including hours of work) into outputs would be appropriate for all. That is, Vernon's research in the First World War related hours to output in a single plant and the IHRB in the Second World War covered a number of plants but all in the munitions industry. In contrast, K&K pooled observations over plants with very different products to glean inferences about the relation between working hours and output.

---

17. In an earlier article, Kossoris (1944) reports preliminary findings on the consequences of long hours of work. Much of the article concerns the increase in absenteeism that follows longer hours. In the plant most discussed, a move from hourly rates of pay to piece rates accompanied the change in hours and, as he recognizes, this frustrates inferences concerning the effect of changes in hours.

18. Although K&K (1948, 37) claim that "work injuries increased disproportionately as hours increased," their information on injuries is marred by differences across plants in record-keeping. When restricted to disabling injuries, there are too few observations to draw confident conclusions.

Hours and Output

With respect to their analysis of output and the length of the working week, K&K investigated situations in which work schedules were altered, and they examined the association between these changes in work schedules and changes in output. They provide information on the hours and days worked before the change and after the change in work schedules, but they do not report levels of output before and after the change in work schedules; they report percentage changes (from "before" to "after") in output. This may well be the consequence of the fact that the output of different plants would be measured in different units, whereas measuring the percentage change in output would allow the comparison of the percentage change in the output of (for example) drugs with the percentage change in the output of airplanes.

In most of their cases, the changes in work schedules took the form of increases in the length of the working week. In some instances (collected at or near the close of the war), the changes in work schedules took the form of reductions in the length of the working week.

There are 92 observations on plants or departments of plants in their analysis of changes in output. Descriptive statistics on the principal variables for these observations are given in table 4.14, where the subscript B indicates the value of the variable before the change in work schedules and the subscript A is the value of the variable after the change in work schedules. The percentage change in weekly output is given by ΔX, weekly hours by H, and days worked per week by D. From the mean and median of ΔX, the central tendency of the change in output is 14 or 15%; the change in weekly hours (that is, $H_A - H_B$) is about seven hours from about 44 hours (or an increase in hours of about 16%).

TABLE 4.14 Descriptive Statistics of the 92 Observations reported by K&K on Changes in Output and Hours and Days of Work

	μ	σ	σ/μ	min	Q_L	M	Q_U	max
ΔX	15.8	15.6	0.99	−7.90	6.45	13.4	20.9	81.2
H_A	51.7	5.57	0.108	35	48	50	57	66
H_B	44.5	5.39	0.121	37.5	40	43.8	48	63
$H_A - H_B$	7.26	6.30	0.87	−13	4.75	8	10	26
D_A	5.63	0.47	0.08	5	5	6	6	6
D_B	5.54	0.50	0.09	5	5	6	6	6

Note: The arithmetic mean is denoted by μ, the standard deviation by σ, the lower quartile by Q_L, the median by M, and the upper quartile by Q_U. The coefficient of variation is σ/μ. The minimum and maximum values of each variable are given under *min* and *max*, respectively.

The subscript A attached to hours (H) and days (D) identifies the values of H and D after the change in work schedules while the subscript B identifies the values of H and D before the change in work schedules. ΔX is the percentage change in output.

As for days of work per week, of the 92 cases, 39 observations were of plants where a change in weekly hours was effected only by changing the hours worked per day and without any change in the number of days worked. There is no instance in the 92 observations of the working week's being increased to seven days. The authors do report on four "clean-cut" cases of regular Sunday work (but values of output and hours for these four are not supplied) and they summarize these cases by writing that (because work on Sunday was at double the pay rate) "Sunday work meant 8 days pay for 6 days output" (4), a comment made by a foreman testifying to Britain's Health of Munition Workers Committee.

With respect to hours of work, their principal conclusion from analyzing these data is "As a rule ... the longer hours yielded higher output—but at a regressive rate. As hours went up, the proportionate return decreased" (1–2). If $\Delta \ln X_j$ represents the percentage change in output in plant j and $\Delta \ln H_j$ the percentage change in weekly hours worked in this plant, then K&K are proposing that, in the regression equation that follows, β_1 is less than unity:

(15) $$\Delta \ln X_j = \beta_0 + \beta_1 \Delta \ln H_j + \varepsilon_{1j}$$

where ε_{1j} is a disturbance incorporating the effects of omitted variables and errors in measuring the change in output. Least-squares estimates of equation (15) are contained in table 4.15 where it is seen that the estimate of β_1 is, indeed, less than unity and significantly so on a one-tailed t test.

To determine the relevance of a nonlinearity in hours in affecting the change in output, the difference in the square of the logarithm of hours, $\Delta (\ln H_j)^2$ was added to the right-hand side of equation (15), as in

(16) $$\Delta \ln X_j = \beta_0 + \beta_1 \Delta \ln H_j + \beta_2 \Delta (\ln H_j)^2 + \varepsilon_{2j}$$

where $\Delta (\ln H_j)^2$ is defined as $\Delta (\ln H_{Aj})^2 - \Delta (\ln H_{Bj})^2$ and the subscripts A and B denote "after" and "before," respectively. As is evident from table 4.15, including this squared term adds virtually nothing to the explanatory fit. Other specifications of the output–hours relation were estimated with K&K's observations,

TABLE 4.15 Least-Squares Estimates of Equations (15) and (16) with K&K's 92 Plants

Equation	Estimated coefficient (and standard error) on		R^2	See
	$\Delta \ln H_j$	$\Delta (\ln H_j)^2$		
15	0.712 (0.154)		0.458	11.56
16	1.150 (1.544)	−6.549 (20.951)	0.460	11.60

Estimates of Production Functions

but what can be done is hampered by not knowing the *level* of output. Because the estimated β_1 is less than unity, the hypothesis of unit elasticity of output with respect to hours is not supported, and the implied marginal product of hours declines with hours throughout for both equations (15) and (16). The law of diminishing returns to hours of work is satisfied.

Hours and Absenteeism

K&K also report on the link between the length of the working week and absenteeism rates. Insofar as long hours induce fatigue and a greater susceptibility to ill-health (as argued in chapter 6), a link is hypothesized between hours of work and absenteeism. The absenteeism rate is the percentage of scheduled work time lost because employees do not show up for work or are not at work for as long as scheduled. There are 176 observations across plants in the absenteeism rate, and the length of work and descriptive statistics on these variables are given in table 4.16. Y is the absenteeism rate expressed as a percentage; H is weekly scheduled

TABLE 4.16 Descriptive Statistics of the 176 Observations reported by K&K and used in Reanalysis of the Effects of the Length of Work on Absenteeism

	μ	min	Q_L	M	Q_U	max	σ	σ/μ	$(\Delta Q)/M$
Y	5.84	0.50	3	4.70	7.80	22.1	3.93	0.673	1.021
H	48.2	35	43	48	54	54	6.57	0.136	0.229
D	5.60	5	5	6	6	6	0.48	0.086	0.167

Note: The absenteeism rate is given by Y (in %).

hours of work; D is days of the working week scheduled. There is little variation in days of work.

K&K also distinguish workplaces by the gender of the workers and by the physical strength needed for the work. In the equations that follow, W_j is a dichotomous variable that takes the value of unity when the workers in workplace j are women only and of zero otherwise, and L_j is another dichotomous variable that takes the value of unity when the work is described as "light" and of zero otherwise. W_j takes the value of unity for 51% of the observations, and L_j is unity for 63% of the observations.

From their own analysis, K&K conclude that "clearly ... the longer the hours, the more scheduled worktime lost through absenteeism" (2). Absenteeism tends to rise when Saturday is added to the five-day working week, it is higher among women than among men, and absenteeism is sometimes lower where the physical demands on the workers are light. An expression of their conclusions is embodied in the following equation:

(17) $$Y_j = \gamma_0 + \gamma_1 H_j + \gamma_2 D_j + \gamma_3 W_j + \gamma_4 L_j + \varepsilon_{3j}$$

where ε_{3j} represents omitted variables and errors in measuring the absenteeism rate.

When estimated by least-squares with heteroskedastic-robust standard errors in parentheses below their associated regression coefficients, the results are as follows:

$$Y_j = \begin{matrix} -6.147 + 0.166 H_j + 0.467 D_j + 4.098 W_j - 1.135 L_j \\ (3.240)(0.044)(0.676)(0.655)(0.497) \end{matrix}$$

with $R^2 = 0.185$ and $see = 3.58$. The estimate of the coefficient on weekly hours implies that if H increases from 48 hours (its mean and median) to 54 hours (its upper quartile value), the absenteeism rate increases by almost 1% (or about 17% of its mean value). Evaluated at mean values of the variables, the elasticity of the absenteeism rate with respect to weekly hours is 1.37.

Absenteeism also rises with days worked, although the estimated coefficient is smaller than its associated standard error. Workplaces with only women workers exhibit a 4% higher absenteeism rate and plants where workers do light work have a 1% lower absenteeism rate. These estimates are consistent with K&K's inferences, although together less than one-fifth of the variation in the absenteeism rate among these plants is removed by this linear combination of the right-hand-side variables.

Observations Collected by Ben Craig on Plywood Mills in the State of Washington

Production functions will now be estimated to observations assembled by Dr. Ben R. Craig while he was on the faculty of Washington State University in Pullman, Washington, in the 1980s. These observations are on individual plywood mills in the state every two years between 1968 and 1986. A survey of these mills was conducted by the state every other year, which accounts for the biennial pattern of the data.

The principal purpose of this project on these plywood mills was to compare the behavior of two types of ownership and management structures.[19] One type was the conventional enterprise owned and managed (often indirectly) by those who supply the firm's financial capital. The other type of ownership structure was the worker co-operative—that is, a firm owned and managed by those who work in the firm. These worker co-operatives in the plywood industry of the Pacific Northwest are believed to have

19. See, for instance, Craig and Pencavel (1992) and Pencavel (2001).

been the most prominent and durable of such labor-managed and labor-owned companies in America during the twentieth century.

Owing to depletion of the old-growth timber forests and the subsequent restrictions on cutting of trees, the prices of the logs from which the plywood is manufactured have risen and the industry has been in decline in the Pacific Northwest during the last forty years. The center of the American plywood industry has shifted to the US South. A number of co-ops have closed in the region, as have conventional mills.

The plywood manufacturing process involves cutting (or "peeling") suitably prepared logs with large lathes, drying the wood, and then bonding the cut wood for home and office construction. It is a highly cyclical industry that tracks the swings in construction activity. In the observations used here, there are 27 conventional mills and 7 co-op mills. The typical mill consisted of about 240 workers, the co-op mills a little larger. Because some mills were not observed in every (even-numbered) year, this is not a balanced data set.

Descriptive statistics on the 170 mill-year observations to be used here are provided in table 4.17. Precise definitions of variables are provided beneath table 4.17. Of the 170 mill-year observations, 134 observations are on the conventional mills and 36 observations are on the co-ops. Descriptive statistics on the co-op mills and conventional mills separately are also given. The conventional mills themselves may be divided into those mills covered by union contracts and a small set of private, nonunion, mills. There are 19 mills and 101 mill-year observations on the unionized plants and 8 mills and 33 mill-year observations on the nonunion plants. The latter are smaller operations, and as statistical tests suggested no significant difference between the unionized and nonunion plants in the first wave of production functions fitted, the work that follows combines the unionized and the nonunion mills.

TABLE 4.17 Descriptive Statistics on the Mill-Year Observations on Washington State's Plywood Industry

All 170 mill-year observations

	H	N	L	M	X
μ	1.87	238.6	459.8	21,925.1	0.10
Min	0.40	8	4.32	100	0.00023
Q_L	1.64	142	239.4	7,781	0.044
M	1.92	227	439.2	19,591	0.11
Q_U	2.01	308	608.6	30,000	0.15
Max	3.50	678	1,328.9	69,774	0.27
σ	0.40	146.8	294.9	15,968.4	0.063
cv	0.214	0.615	0.641	0.728	0.63
$(\Delta Q)/M$	0.193	0.731	0.841	1.134	0.964

134 mill-year observations on conventional mills

	H	N	L	M	X
μ	1.79	230	426.7	21,712.7	0.094
min	0.40	8	4.32	100	0.00023
Q_L	1.60	100	183.5	7,000	0.028
M	1.88	206	390.5	17,000	0.092
Q_U	2	306	590.0	34,000	0.15
max	3.50	678	1,328.9	69,774	0.27
σ	0.39	159.4	310.1	17,060	0.067
cv	0.205	0.693	0.727	0.786	0.713
$(\Delta Q)/M$	0.213	1.0	1.041	1.588	1.326

36 mill-year observations on worker co-op mills

	H	N	L	M	X
μ	2.15	270.7	583.1	22,715.8	0.13
min	1.48	136	239.4	2,998	0.073

(*continued*)

DIMINISHING RETURNS AT WORK

TABLE 4.17 continued

	H	N	L	M	X
Q_L	2	232	481.8	16,022.5	0.11
M	2.06	250	565.4	23,000	0.13
Q_U	2.34	313	625.9	28,569	0.16
max	2.92	466	1,077.5	59,672	0.021
σ	0.28	78.6	186.5	11,156.1	0.033
cv	0.130	0.290	0.320	0.491	0.254
$(\Delta Q)/M$	0.165	0.324	0.255	0.546	0.385

Note: H denotes annual hours worked per worker measured in thousands of hours. N is the number of workers employed in the mill averaged over the weeks in the year. L is worker-hours— i.e., the product of H and N. M is the quantity of raw material logs (measured in thousands of feet) used by the mill in the year. X is annual output in square feet of softwood plywood and veneer. Plywood and veneer are aggregated using region-specific current prices and then deflated by a plywood producer price index.

The arithmetic mean is μ. The lowest value is *min* and the highest value is *max*. Q_L is the lower quartile (or 25th percentile) and Q_U is the upper quartile (or 75th percentile). M is the median. The standard deviation is given by σ and the coefficient of variation by *cv*.

$\Delta Q/M$ is $(QU-QL)/M$

All inputs and output are measured as annual values. Annual hours per worker are computed as the number of days of operation over the year times average hours of work per day. These are then expressed in units of thousands. The central tendency of hours is approximately 1.9 thousand hours.

Variations in hours and employment are noticeably smaller for the co-ops than for the conventional mills. This is because the co-ops' principal response to price shocks took the form of altering the payments to their worker-members, whereas the conventional mills tended to adjust input and output levels and to leave wages unchanged. Each mill (both the conventional and the co-op) faces predetermined prices of plywood output and of log inputs. These prices are volatile. Each conventional mill covered by a collective bargaining contract faced a region-wide negotiated wage rate that allowed for differences across mills in hourly earnings.

In earlier research (Craig and Pencavel 1995), the observations on annual plywood output and annual values of inputs (including logs and the number of worker-hours) were used to compare the productivity of the co-op mills with that of the conventional mills. Here, attention is first focused on the manner in which hours of work enter the production function. Invoking the specifications used in the equations estimated earlier, the following stochastic equations were fitted to the 170 mill-year observations on the plywood industry:

(18) $\ln X_{jt} = \beta_0 + \beta_1 \ln H_{jt} + \eta_1 \ln N_{jt} + \eta_2 \ln M_{jt} + \rho_1 T_t + \omega_j + \varepsilon_{jt}$

(19) $\ln X_{jt} = \beta_0 + \beta_2 \ln H_{jt} + \beta_3 (\ln H_{jt})^2 + \eta_3 \ln N_{jt} + \eta_4 \ln M_{jt} + \rho_2 T_t + \omega_j + \varepsilon_{jt}$

(20) $\ln X_{jt} = \alpha_0 + \alpha_1 H_{jt} + \eta_5 \ln N_{jt} + \eta_6 \ln M_{jt} + \rho_3 T_t + \omega_j + \varepsilon_{jt}$

(21) $\ln X_{jt} = \alpha_0 + \alpha_1 H_{jt} + \alpha_2 (H_{jt})^2 + \eta_7 \ln N_{jt} + \eta_8 \ln M_{jt} + \rho_4 T_t + \omega_j + \varepsilon_{jt}$

In these equations, X_{jt} represents the output of plywood in mill j in year t, H_{jt} is the annual hours worked per worker in mill j in year t, N_{jt} is the employment of workers in mill j in year t, and M_{jt} is the annual input of raw material logs by mill j in year t. Although some data were collected on lathes, a comprehensive measure of physical capital was not available. The assumption underlying the four equations above is that the principal elements of physical capital, such as the equipment and capacity of the mill, differed across mills but changed smoothly (perhaps depreciated) over time and that these differences and changes are embodied in fixed plant effects (given by ω_j) and a yearly time trend (given by T_j). Equations (18) and (19) are slight modifications of the specification inspired by Cobb and Douglas, as in equations (CD-1) and (CD-2). Equations (20) and (21) adapt the Gompertz-inspired equations (GZ-1) and (GZ-2).

The least-squares estimates of these equations are contained in table 4.18 and, once again, the quadratic terms in hours, contained in equations (19) and (21), contribute significantly to accounting for variations in the logarithm of output. The output–hours relation implied by the estimates of these two equations are shown in figure 15.[20] These two equations differ in their implications for the slopes of the output–hours relation: the estimates of equation (19) imply output is maximized when annual hours per worker reach 3,326, a number barely within the range of observed values of H, while those of equation (21) imply output reaches a maximum at 2,500 hours. Differences between the estimates of equations (19) and (21) are evident from the implied marginal products of hours portrayed in figure 16.

The estimates of equation (18) in table 4.18 imply that a 10% reduction in hours reduces output by 6%. Hence, the hypothesis of unit returns to hours is not sustained. The estimates of equation (19) imply that the elasticity of output with respect to

---

20. In drawing the implied output–hours relations in figures 15 and 16, the values of N and M are held fixed at their approximate mean values.

TABLE 4.18 Least-Squares Estimates of Equations (18), (19), (20), (21), and (22) Fitted to 170 Mill-Year Observations on the Plywood Industry

Right-hand-side variable ⇓	Equation (18)	Equation (19)	Equation (20)	Equation (21)	Equation (22)
$\ln H_{jt}$	0.661 (0.131)	0.899 (0.168)			0.847 (0.186)
$(\ln H_{jt})^2$		−0.374 (0.137)			−0.349 (0.150)
H_{jt}			0.314 (0.087)	1.517 (0.370)	
$(H_{jt})^2$				−0.304 (0.083)	
$\ln N_{jt}$	0.312 (0.128)	0.312 (0.125)	0.327 (0.130)	0.316 (0.129)	0.998 (0.885)
$(\ln N_{jt})^2$					−0.070 (0.088)
$\ln M_{jt}$	0.510 (0.072)	0.488 (0.066)	0.564 (0.79)	0.488 (0.075)	0.511 (0.070)
T_t	0.011 (0.005)	0.008 (0.005)	0.012 (0.005)	0.010 (0.005)	0.008 (0.005)
Goodness of fit statistics					
R^2	0.957	0.959	0.953	0.957	0.959
See	0.260	0.255	0.271	0.258	0.255

Note: Estimated standard errors in parentheses are heteroskedastic-robust. see is the standard error of estimate of the fitted equation.

(18) $\ln X_{jt} = \beta_0 + \beta_1 \ln H_{jt} + \eta_1 \ln N_{jt} + \eta_2 \ln M_{jt} + \rho_1 T_t + \omega_j + \varepsilon_{jt}$

(19) $\ln X_{jt} = \beta_0 + \beta_2 \ln H_{jt} + \beta_3 (\ln H_{jt})^2 + \eta_3 \ln N_{jt} + \eta_4 \ln M_{jt} + \rho_2 T_t + \omega_j + \varepsilon_{jt}$

(20) $\ln X_{jt} = \alpha_0 + \alpha_1 H_{jt} + \eta_5 \ln N_{jt} + \eta_6 \ln M_{jt} + \rho_3 T_t + \omega_j + \varepsilon_{jt}$

(21) $\ln X_{jt} = \alpha_0 + \alpha_1 H_{jt} + \alpha_2 (H_{jt})^2 + \eta_7 \ln N_{jt} + \eta_8 \ln M_{jt} + \rho_4 T_t + \omega_j + \varepsilon_{jt}$

(22) $\ln X_{jt} = \beta_0 + \beta_2 \ln H_{jt} + \beta_3 (\ln H_{jt})^2 + \eta_3 \ln N_{jt} + \eta_5 (\ln N_{jt})^2 + \eta_4 \ln M_{jt} + \rho_5 T_t + \omega_j + \varepsilon_{jt}$

DIMINISHING RETURNS AT WORK

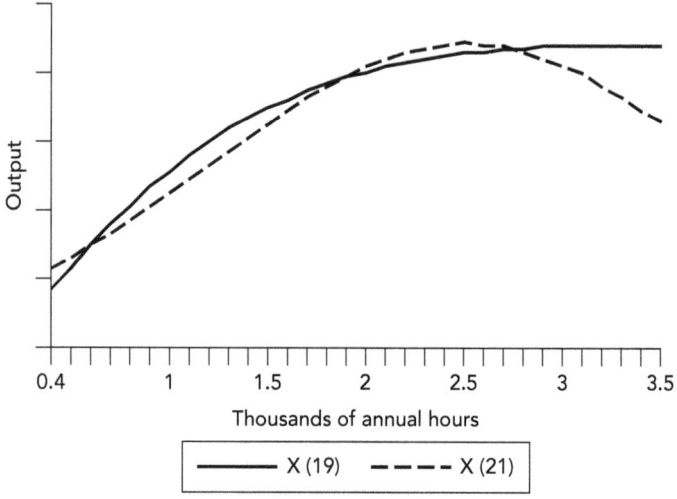

FIGURE 15 The Output–Hours Relation Implied by the Estimates of Equations (19) and (21) Fitted to the 170 Mill-Year Observations on the Plywood Industry

(19) $\ln(X_{jt}) = \beta_0 + \beta_2 \ln(H_{jt}) + \beta_3 \left[\ln(H_{jt}) \right]^2 + \eta_3 \ln(N_{jt}) + \eta_4 \ln(M_{jt}) + \rho_2 T_t + \omega_j + \varepsilon_{jt}$

(21) $\ln(X_{jt}) = \alpha_0 + \alpha_1 H_{jt} + \alpha_2 (H_{jt})^2 + \eta_7 \ln(N_{jt}) + \eta_8 \ln(M_{jt}) + \rho_4 T_t + \omega_j + \varepsilon_{jt}$

hours, $\mathscr{E}(X, H)$ is not a constant but falls as hours lengthen: inspection of figure 17 that uses the estimates of equation (19) to graph $\mathscr{E}(X, H)$ shows that this effect is not significantly different from unity between 700 annual hours and 1,000 annual hours, but for hours longer than 1,000 hours, this proportionate effect is significantly below unity. Most observations on hours in these plywood data are beyond 1,000 hours, so the hypothesis of constant returns to hours is not an appropriate description of these observations. The decline in $\mathscr{E}(X, H)$ as hours increase confirms the textbook movements in the marginal product of an input (here, hours of work) relative to the average product of that input, as in figure 2.

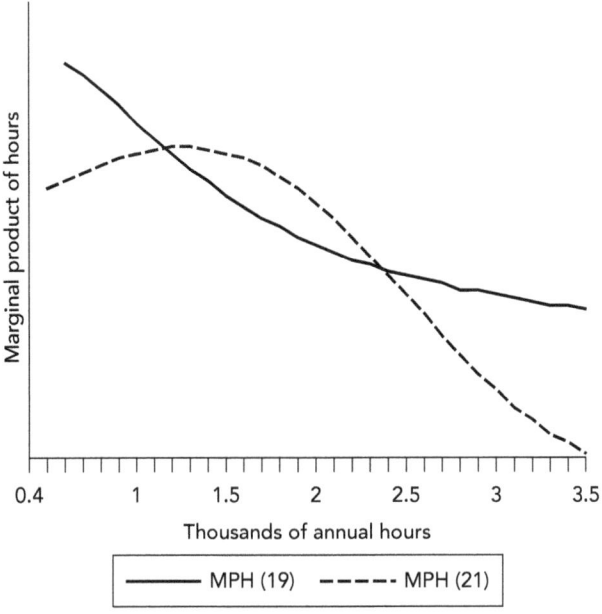

FIGURE 16 The Marginal Product of Hours Implied by the Estimates of Equations (19) and (21) Fitted to the 170 mill-year observations on the Plywood Industry

(19) $\ln(X_{jt}) = \beta_0 + \beta_2 \ln(H_{jt}) + \beta_3 \left[\ln(H_{jt})\right]^2 + \eta_3 \ln(N_{jt}) + \eta_4 \ln(M_{jt}) + \rho_2 T_t + \omega_j + \varepsilon_{jt}$

(21) $\ln(X_{jt}) = \alpha_0 + \alpha_1 H_{jt} + \alpha_2 (H_{jt})^2 + \eta_7 \ln(N_{jt}) + \eta_8 \ln(M_{jt}) + \rho_4 T_t + \omega_j + \varepsilon_{jt}$

What about the "lump of labor" proposition—that, in measuring the input of labor in production, hours per worker and the number of workers may be combined into a single variable, worker-hours? If this proposition is correct, in equation (18), the elasticity of output with respect to hours (β_1) will equal the elasticity of output with respect to employment (η_1). Inspection of the least-squares estimates of equation (18) in table 4.18 indicates that the point estimate of β_1 is twice that of η_1. However, the estimated standard errors are such that, by conventional criteria,

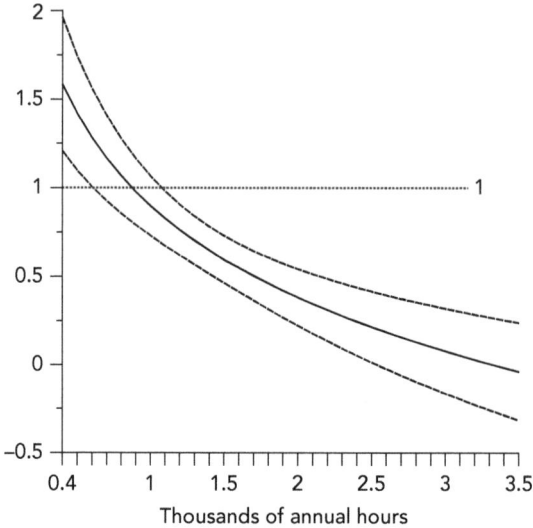

FIGURE 17 Estimated Values of $\mathcal{E}(X, H)$ as Implied by Equation (19) Fitted to the Plywood Industry.

The dotted horizontal line indicates a value of $\mathcal{E}(X, H)$ of unity. The dashed curves denote 95 percent confidence intervals around $\mathcal{E}(X, H)$.

the hypothesis that $\eta_1 = \beta_1$ cannot be rejected. Yet, the standard error of estimate of the equation suggests the specification of equation (19) provides a superior fit to that of equation (18), and the elasticity of output with respect to hours in equation (22) is not a constant, as shown in Figure 17.

According to the estimates of equations (18) and (19) in table 4.18, the elasticity of output with respect to employment is 0.312. The estimates of equation (19) imply that the elasticity of output with respect to hours equals 0.312 when hours of work are 2.19 thousand per year, which is approximately the mean value of hours in these data. The conclusion might be that, over some hours, the proposition that the two elasticities are equal will describe observations well, but not at all hours, and it would be prudent not to assume that equality but to test for it.

A case has been made for hours of work to enter the logarithmic production function in a nonlinear fashion. Is there empirical support for employment also to enter in this form? To answer this, consider the following equation:

(22) $\quad \ln X_{jt} = \beta_0 + \beta_2 \ln H_{jt} + \beta_3 (\ln H_{jt})^2 + \eta_3 \ln N_{jt}$
$\quad\quad\quad + \eta_5 (\ln N_{jt})^2 + \eta_4 \ln M_{jt} + \rho_5 T_t + \omega_j + \varepsilon_{jt}$

where a quadratic term in the logarithm of employment has been included. The least-squares estimates of equation (22) are contained in table 4.18. The point estimate of η_5 is smaller than its standard error, and there is no improvement in the fit over that of equation (19). The estimated effects of hours of work are unaltered. There is meager empirical support for the hypothesis that employment enters the production function in a manner synonymous with hours of work.

To consider the lump of labor proposition further, if it were true that the input of labor is suitably measured by worker-hours, if equation (18) were a correct specification of the production function with $\beta_1 = \eta_1$, and if the conventional mills maximized their net returns given product and input prices, then the mills' input demand function for hours would have the same implications for the effects of changes or differences in the mills' price environment as the mills' input demand function for workers.[21]

To be specific, if a firm's production function is the conventional Cobb-Douglas production function, if the firm selects its

---

21. The characterization of these mills as price-takers in input and output markets conforms closely to their situation. The price of plywood is determined in international markets. As for their use of logs, most were grown on public lands and then purchased by the mills in public auctions. Most of the conventional mills were covered by a region-wide collectively bargained contract that allowed some but little variation in wages across mills at a given time.

inputs to maximize its profits, and if it faces predetermined prices for its inputs and output, its input demand functions are linear in the logarithms in these prices. The hours of work equation and the employment equation would take the following form:

(23) $\ln H_{jt} = a_0 + a_1 \ln w_{jt} + a_2 \ln r_{jt} + a_3 \ln p_{jt} + \sigma_1 T_t + \omega_j + \varepsilon_{jt}$

(24) $\ln N_{jt} = b_0 + b_1 \ln w_{jt} + b_2 \ln r_{jt} + b_3 \ln p_{jt} + \sigma_2 T_t + \omega_j + \varepsilon_{jt}$

where w denotes hourly earnings, r the unit price of logs, and p the price of plywood output. As before, T_t is a yearly time trend, ω_j is a fixed effect for each mill, and ε_{jt} is a stochastic term assumed to have properties amenable to least-squares estimation.

The proposition that, for the specification of labor in the production function, hours per worker and the number of workers may be combined into a single variable, worker-hours, implies that, under the circumstances given, $a_1 = b_1$, $a_2 = b_2$, and $a_3 = b_3$. The least-squares estimates of equations (23) and (24) fitted to the 134 mill-year observations on the capital-owned and capital-managed mills are given in table 4.19.

Inspection of table 4.19 reveals that the parameter estimates in the hours equation resemble those in the employment equation, in that the *signs* of the coefficients on each right-hand-side variable in the hours equation are the same as those in the employment equation. Nevertheless, a formal F test rejects the null hypothesis that the coefficients of the two equations are the same. The estimates imply that employment is more sensitive to changes in prices and wages than is hours per worker. This is to be expected if the conventional Cobb-Douglas production misspecifies the manner in which hours of work affect output. If the correct specification of the production function is equation (19), then equation (23) is not the implied demand for hours of work, but only an approximation of it.

Earlier research investigating differences in production functions between the conventional mills and co-operative mills

TABLE 4.19 Least-Squares Estimates of Equations (23) and (24) Fitted to 134 Mill-Year Observations on the Conventional Mills in the Plywood Industry

Right-hand-side variable ⇓	Left-hand-side variable	
	$\ln H_{jt}$	$\ln N_{jt}$
$\ln w_{jt}$	−0.275 (0.228)	−0.336 (0.154)
$\ln r_{jt}$	−0.121 (0.105)	−0.195 (0.135)
$\ln p_{jt}$	0.425 (0.159)	0.625 (0.148)
T_t	−0.008 (0.004)	−0.011 (0.006)
Goodness of fit statistics		
R^2	0.497	0.956
see	0.214	0.213

Note: Estimated standard errors in parentheses are heteroskedastic-robust. *see* is the standard error of estimate of the fitted equation.

(23) $\quad \ln H_{jt} = a_0 + a_1 \ln w_{jt} + a_2 \ln r_{jt} + a_3 \ln p_{jt} + \sigma_1 T_t + \omega_j + \varepsilon_{jt}$

(24) $\quad \ln N_{jt} = b_0 + b_1 \ln w_{jt} + b_2 \ln r_{jt} + b_3 \ln p_{jt} + \sigma_2 T_t + \omega_j + \varepsilon_{jt}$

found "there is not much to distinguish these types of firms in terms of overall production efficiency. What differences we have found imply that co-ops are more efficient than the principal conventional firms by between 6 and 14 per cent" (Craig and Pencavel 1995, 158). This conclusion rested on a conventional Cobb-Douglas production function in which hours of work were combined with employment to form worker-hours. This book questions this form of the production function and argues that this specification should be subject to prior tests: first, to see whether worker-hours should be unbundled into their distinct components (employment and hours of work); and second, to determine whether the effect of hours of work on output declines as hours lengthen.

To address the issue of the relative productivity of the conventional mills and the co-operatives in the context of the new specification of the production function, we replace the mill fixed effects in equations (19) and (21) with a dichotomous variable C_{jt} that takes the value of unity if plant j in year t took the co-operative form and of zero otherwise.[22] This yields the following two specifications:

(25) $\quad \ln X_{jt} = \beta_0 + \delta_1 C_{jt} + \beta_2 \ln H_{jt} + \beta_3 (\ln H_{jt})^2 + \eta_3 \ln N_{jt}$
$\quad\quad\quad + \eta_4 \ln M_{jt} + \rho_2 T_t + \varepsilon_{jt}$

(26) $\quad \ln X_{jt} = \alpha_0 + \delta_2 C_{jt} + \alpha_1 H_{jt} + \alpha_2 (H_{jt})^2 + \eta_7 \ln N_{jt}$
$\quad\quad\quad + \eta_8 \ln M_{jt} + \rho_4 T_t + \varepsilon_{jt}$

Least-squares estimates of these two equations are reported in table 4.20. The estimates imply that the co-operative mills produced between 19 and 21% more output, holding constant hours of work, employment, and the input of logs.[23] The previous conclusions that the cooperative mills were more productive than the conventional mills based on a different expression for the production function are sustained. Of course, superior knowledge of one of a firm's activities—in this case, the production technology—does not ensure commercial success. The implications of the comparative higher productivity of the worker co-ops are taken up in Pencavel (2001).

---

22. C_{jt} has a t subscript because, over this period, two mills changed their organizational form. Peninsula Plywood was converted from a co-op to a unionized mill between 1970 and 1972: it is observed as a co-op in 1968 and 1970 and as a unionized conventional mill between 1972 and 1986. Anacortes is observed as a unionized conventional mill in 1968, 1970, 1972, 1974, 1976, 1978, and 1980. After conversion to a co-op, it is observed as a co-op in 1984 and 1986.

23. That is, from the estimates of δ_1 and δ_2 in table 4.20, the proportional difference in output between the two types of enterprises is $exp(0.187)-1=1.206-1=0.206$ in the case of equation (25) and $exp(0.177)-1=1.194-1=0.194$ for equation (26).

TABLE 4.20 Least-Squares Estimates of Equations (25), and (26) Fitted to 170 Mill-Year Observations on the Plywood Industry

$$(25)\ \ln X_{jt} = \beta_0 + \delta_1 C_{jt} + \beta_2 \ln H_{jt} + \beta_3 (\ln H_{jt})^2 + \eta_3 \ln N_{jt} + \eta_4 \ln M_{jt} + \rho_2 T_t + \varepsilon_{jt}$$

$$(26)\ \ln X_{jt} = \alpha_0 + \delta_2 C_{jt} + \alpha_1 H_{jt} + \alpha_2 (H_{jt})^2 + \eta_7 \ln N_{jt} + \eta_8 \ln M_{jt} + \rho_4 T_t + \varepsilon_{jt}$$

Right-hand-side variable ⇓	Equation (25)	Equation (26)
C_{jt}	0.187 (0.076)	0.177 (0.074)
$\ln(H_{jt})$	0.854 (0.182)	
$[\ln(H_{jt})]^2$	−0.057 (0.170)	
H_{jt}		1.490 (0.353)
$(H_{jt})^2$		−0.266 (0.075)
$\ln(N_{jt})$	0.592 (0.053)	0.585 (0.054)
$\ln(M_{jt})$	0.501 (0.053)	0.493 (0.059)
T_t	0.008 (0.007)	0.008 (0.007)
Goodness of fit statistics		
R^2	0.824	0.825
see	0.473	0.471

Note: Estimated standard errors in parentheses are heteroskedastic-robust. see is the standard error of estimate of the fitted equation.

Some Conclusions About Hours and Output of Workers from These Four Cases

Results from estimating the output–hours relations specified in chapter 3 to these foregoing four sets of observations have been presented—two sets from Britain and two sets from America. In one of these cases, that undertaken by Kossoris and Kohler, inferences are hampered by the manner in which the data have been reported and by the fact that the observations do not describe a single production technology. In this case, all that was done with respect to the output–hours link was to confirm the authors' own conclusions: that, in these data, the elasticity of output with respect to hours was less than unity (or, equivalently, the estimate of θ in equation (3) is negative, implying that work effort per hour falls as hours of work lengthen). Kossoris and Kohler did report that absenteeism rates rose significantly as hours increased, and this is relevant to the discussion on health and hours in chapter 6. This finding was confirmed.

The other three sets of observations—Horace Vernon's data on British munition workers in a plant during the First World War, the IRHB's data from six British munition plants during the Second World War, and Ben Craig's data on thirty-four plywood mills in Washington State between 1968 and 1986—allow the estimation of the production functions specified in chapter 3. The estimates imply that the efficacy of workers in generating an output declines as the length of their work increases, and the hypothesis of constant returns to working hours is not supported by these results. In these cases, the estimated elasticity of output with respect to hours, $\mathcal{E}(X, H)$, declines as hours increase, which is consistent with the variation in the ratio of the marginal product of hours to the average product of hours pictured in figure 2 after H_0.

Consistently, these estimates imply that effort per hour declines as hours lengthen, and effective hours follow a positive concave path with respect to observed hours. Hence, an additional hour when workers are working "long" hours is less productive than an additional hour when workers are working "short" hours.

Typically, as a description of the proportional variations in output, the estimates of Cobb and Douglas' augmented and Gompertz' augmented production functions represent a significant improvement over their nonaugmented versions: adding a quadratic term in hours or a quadratic term in the logarithm of hours results in a statistically significant increase in the variance in the logarithm of output removed by variations in the linear combination of right-hand-side variables. This is a statistical criterion. Does the addition of these quadratic terms in hours constitute much in the way of economic significance?

One way to address this is to compare the output implied by the estimated nonaugmented specification with the output implied by the estimated augmented specification. If there is little difference between the implied output of the two specifications, the principle of parsimony would suggest a preference for the nonaugmented version. To this effect, consider these definitions, first for the Cobb-Douglas function:

1. $[X_O(H_k)]^{CD}$ is the output implied at hours H_k by the estimates of Cobb and Douglas's nonaugmented or conventional production function, equation (CD-1), where the subscript O on output indicates that this is the specification that omits the quadratic term in the logarithm of hours.
2. $[X_W(H_k)]^{CD}$ is the output implied at hours H_k by the estimates of Cobb and Douglas's augmented production function, equation (CD-2), where the subscript W on output denotes the specification with the $(\ln H_{jt})^2$ term.

(27) $\quad [\Delta X_D(H_k)]^{CD} = 100\{[X_W(H_k)]^{CD} - [X_O(H_k)^{CD}]\}/\mu_X$

$[\Delta X_D(H_k)]^{CD}$ is the percentage difference (hence, the subscript D on X) between the augmented and nonaugmented Cobb-Douglas functions in the implied output at H_k hours expressed as a fraction of the mean value of output μ_X.

Analogously for the estimates of Gompertz' *(GZ)* specification:

3. $[X_O(H_k)]^{GZ}$ is the output implied at hours H_K by the estimates of Gompertz' nonaugmented production function, equation (GZ-1), where the subscript O on output indicates that this is the specification that *o*mits the quadratic term in hours

4. $[X_W(H_k)]^{GZ}$ is the output implied at hours H_k by the estimates of Gompertz' augmented production function, equation (GZ-2), where the subscript W on output indicates that this is the specification *w*ith the quadratic term in hours.

Now, we compare the output implied by the estimates that include the square of hours of work with the output implied by the estimates that omit the square of hours:

$$(28) \quad [\Delta X_D(H_k)]^{GZ} = 100\{[X_W(H_k)]^{GZ} - [X_o(H_k)]^{GZ}\}/\mu_X$$

$[\Delta X_D(H_k)]^{GZ}$ is the percentage *d*ifference (hence the D subscript on X) between the augmented and nonaugmented Gompertz functions in the implied output at H_k hours expressed as a fraction of the mean value of output μ_X.

The values of $[\Delta X_D(H_k)]^{CD}$ and of $[\Delta X_D(H_k)]^{GZ}$ provide cardinal indicators of the degree to which, at hours H_k, the output implied by the estimates of the augmented production function diverge from the output implied by the estimates of the production function without the additional term of hours squared. Equation (27) constructs this indicator for Cobb and Douglas' logarithmic functional form, and equation (28) uses the estimates of Gompertz' semi-logarithmic functional form at different hours. Figures 18, 19 and 20 report the values of $[\Delta X_D(H_k)]^{CD}$ and of $[\Delta X_D(H_k)]^{GZ}$ for these three bodies of observations.

Most of the values of these indicators are negative, implying that the conventional specifications—both that of Cobb and Douglas and that of Gompertz tend to overpredict output relative to the augmented Cobb-Douglas and augmented Gompertz functions. This relative overprediction is greatest at relatively short hours and at relatively long hours. Over the middle range of hours, the difference in implied output between the augmented and nonaugmented expressions is smaller. The values of Gompertz' differences, equation (28), vary more with hours than

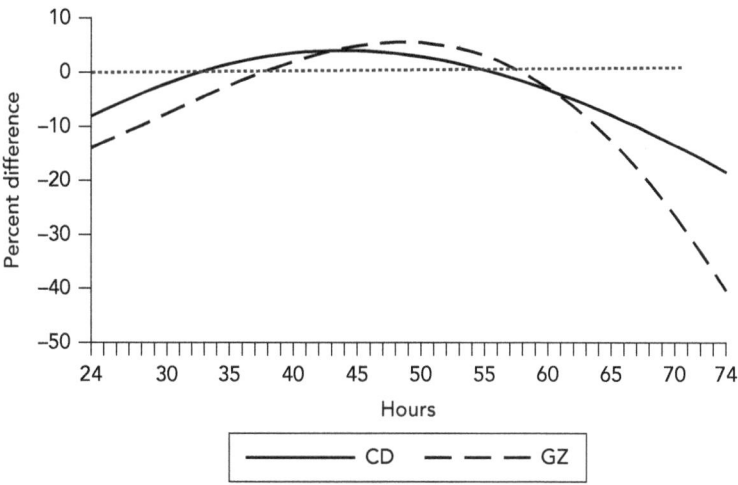

FIGURE 18 The Percentage Difference in Output Implied by the Augmented and Nonaugmented Production Functions of Cobb and Douglas and of Gompertz Fitted to Vernon's Observations on Munitions Workers.

CD corresponds to Cobb and Douglas's production function and GZ denotes Gompertz's production function.

DIMINISHING RETURNS AT WORK

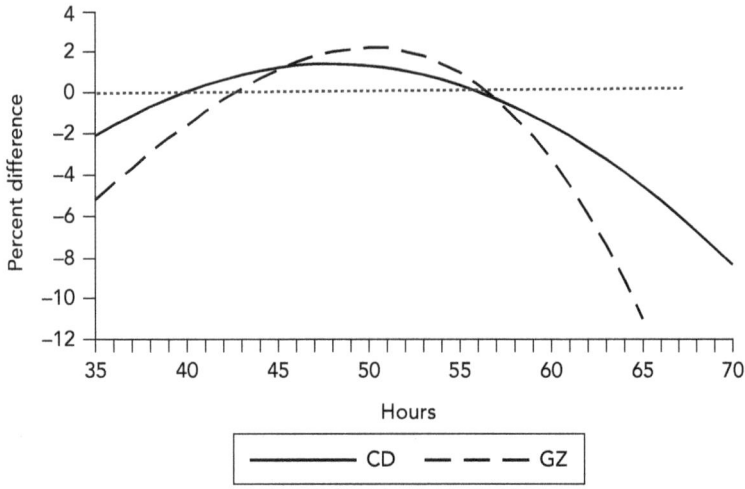

FIGURE 19 The Percentage Difference in Output Implied by the Augmented and Nonaugmented Production Functions of Cobb and Douglas and of Gompertz Fitted to the Industrial Health Research Board's Observations on Munitions Plants.

CD corresponds to Cobb and Douglas's production function and GZ denotes Gompertz's production function.

those of Cobb and Douglas' equation (27). If the nonaugmented production functions were used to infer the output consequences of "long" hours (say, more than 55 hours per week), the additional output implied would considerably overstate the additional output implied by the augmented production functions.

Does it matter whether one uses a conventional specification for the production function or an expression that allows for hours to enter as a higher order term? The answer would seem to be "it depends." If the range of hours in the observations under analysis is deemed to be narrow, the consequences may well be second-order and the conservation of degrees of freedom may incline a

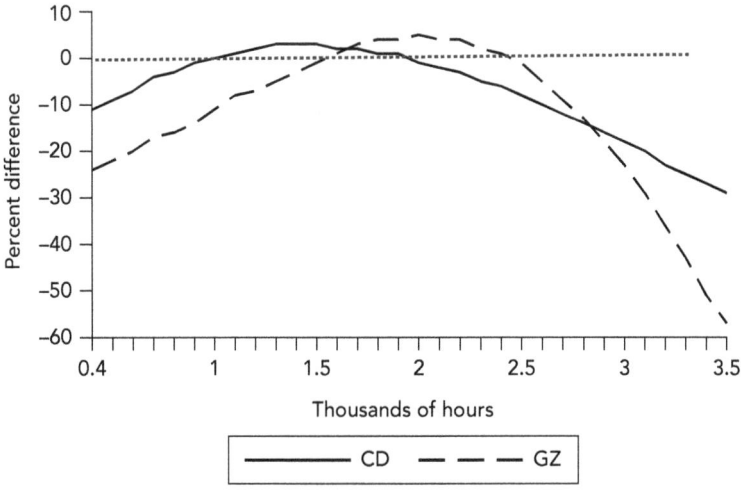

FIGURE 20 The Percentage Difference in Output Implied by the Augmented and Nonaugmented Production Functions of Cobb and Douglas and of Gompertz Fitted to Ben Craig's Observations on Plywood Mills.

CD corresponds to Cobb and Douglas's production function and GZ denotes Gompertz's production function.

choice in favor of the nonaugmented specifications. Where some observations are at short hours or at long hours or at both, a specification that allows for more curvature in the output–hours relation (such as by entering hours in a nonlinear manner) ought to be investigated.

For all the observations used here to fit production functions, the marginal product of hours declines as hours increase, so the principle of diminishing returns to hours is confirmed. The hypothesis of *unit returns to hours* is rejected as a description over the whole range of hours worked. For particular segments of working hours, unit returns is a satisfactory approximation. However, if a specification is sought that is faithful to the textbook portrayal of the relative variation of the marginal product to the average product of an input (as in figure 2), then the augmented production functions are needed. Certainly, the conventional Cobb-Douglas function will not do unless a narrow range of hours is observed.

5

Further Implications of the Augmented Production Functions

The production functions estimated in chapter 4 examined groups of workers who work different hours in various weeks (or years). Variations in their average hours were linked to variations in their output, and in so doing, the textbook relation between the marginal product of hours (as given by the height of *MPH* in figure 2) and hours was affirmed: after a certain level of hours, holding other inputs constant, the *MPH* declined as hours increased.

Part-Time and Full-Time Workers

Consider now the productivity of groups of workers, some of whom routinely work "shorter" hours (call them part-time workers) and some who routinely work "longer" hours (full-time workers). If a firm employs both part-time workers and full-time workers, will the firm's marginal product of nominal hours, *MPH*, increase or decrease if the share of part-time employees in its employment rises? According to the *MPH* curve in figure 2, if both groups of workers are working more than H_O hours, the part-time workers will have a higher *MPH* and, therefore, a firm that increased its share of part-time workers will experience a higher *MPH*. If the part-time workers are working fewer than H_O hours and the full-time workers are working more than H_O hours, the group with the higher *MPH* cannot be determined a priori. What does empirical work suggest?

Consider a population of pharmacies and suppose the output of each pharmacy is measured by the number of medical prescriptions filled in a year. In a cross-sectional study of 235 pharmacies in the Netherlands in 2007, Künn-Nelen et al. (2013) distinguish the part-time workers who were working fewer than 24 weekly hours (three working days) from the full-time workers who were working 36 hours. They also identify a third group of "other employees" (who include administrators and cleaners whose hours are not provided). The researchers report that, holding constant a number of other variables, increases in the share of part-time employees has a larger (positive) effect on output than increases in the share of full-time employees. That is, interpreting the effect on output of an increase in the employment share of part-time workers as the *MPH* of part-time workers, and that of full-time workers as the *MPH* of full-time workers, additional hours of work performed by people working "short" hours raises the typical firm's *MPH* more than additional hours done by people working "long" hours.

Note that some investigators examine variations in output per hour, sometimes referred to as *productivity* or *labor productivity*. As mentioned in chapter 3, when $\ln(H_{jt})$ is subtracted from both sides of equations (CD-1) and (CD-2), an expression for output per hour is derived, such as (CD-1)*. If the law of diminishing returns operates for hours of work, increases in the logarithm of hours should have depressing effects on productivity (output per hour).

Further Implications of the Augmented Production Functions

In an examination of 1,430 Belgian firms over twelve years, from 1999 to 2010, Garnero et al. (2014) distinguish not only between full-time and part-time workers but also between part-time workers working shorter hours and part-time workers working longer hours. The threshold defining full-time work is 35 hours per week. The threshold separating the shorter part-time workers from the longer part-time workers differs across fitted equations. When this threshold is set at 25 weekly hours, in an equation accounting for variations in the logarithm of value added per hour (that is, the logarithm of the value of *average output*) and including a large number of covariates, the estimated coefficient attached to the share of total hours worked by those working fewer than 25 hours would not be judged as statistically different from the reference category of full-time work, but the estimated coefficient on the share of total hours worked by the longer part-time workers (those working 25 to 35 hours per week) is significantly larger than either of the other two categories.

This is what is implied by the specifications in chapter 3: output per hour is a nonlinear function of hours, with in this particular case the maximum of output per hour occurring between 25 and 35 weekly hours. A similar result was derived with the munitions workers studied by Vernon: in figure 9, the average product of hours peaks at about 40 weekly hours. Also refer to the APH curve in figure 2.

Collewet and Sauermann (2017) analyzed 332 workers in a call center in the Netherlands between mid-2008 and January 2010. Their productivity measure is the average time taken (over a day or week) by an individual worker to answer and resolve each call. With 33,123 worker-day observations and a rich set of covariates, they estimate with precision the presence of diminishing returns to hours. Even though these are part-time workers, the authors are able to conclude that "if working hours increase by one percent, the number of calls handled will increase by only 0.9 percent." (22) In short, diminishing returns to hours operates even among part-time workers.

This is also the finding of Brachet et al. (2012), who examined the response time and other indicators of the effectiveness of 2,381 paramedics in Mississippi between 2001 and 2005. Controlling for a number of variables, they report declining effectiveness as the duration of the worker's shift lengthens. They also compare the performance of medics on 12-hour shifts with the performance of these *same* medics on 24-hour shifts, and they find that their response time tends to be longer when working on the 24-hour shift, especially in the early hours of the morning. They write "fatigue plays a role in the decline of the performance of paramedics at the end of long shifts" and "greater attention to the design of work schedules may entail benefits to patients relying on emergency medical services." (244)

Overtime Work

Related to this, in an inquiry of nineteen two-digit German manufacturing industries observed biennially between 1968 and 1978 (380 industry-year observations in all), Hart and McGregor (1988) fit conventional Cobb-Douglas production functions that allow for different effects of hours of work per worker. They "firmly reject" the specification in which hours per worker and the number of workers are combined into one variable, worker-hours. They also distinguish between the hours of standard time worked and the hours of overtime worked. Their point estimate of the productivity of overtime hours is lower than that of normal hours.

Similarly, Shepard and Clifton (2000) estimate production functions for eighteen US manufacturing industries over time, from 1956 to 1991, and conclude that a 10% increase in overtime hours is associated with a decrease of 2 to 4% in labor productivity (output per hour) for most industries.

There has long been the claim that overtime hours tend to be relatively unproductive work and these studies confirm this.

Macroeconomics

In this book, emphasis has been placed on estimating the output–hours relationship at the microeconomic level, at which a well-defined production function is most plausible. It is heroic to suppose that the same production function can meaningfully describe an aggregate of industries. Others are more courageous and have estimated equations resembling these production functions at the macroeconomic level. For example, in an analysis of between fourteen and twenty-five countries, Bourlès and Cette (2005) have related output per hour to average annual hours worked at the aggregate level between 1992 and 2001. The equation is estimated in first-difference form. They find that longer hours are associated with lower output per hour. Nonlinearities along the lines of equation (CD-2) are not investigated.

6

Hours of Work, Health, and Well-Being

The research reported in chapters 4 and 5 concerns the nature of the link between hours of work and the production of a marketable commodity. The research described in this chapter addresses the hypothesis that long working hours damage the health of workers. This hypothesis is plausible if ill-health is envisaged as jointly "produced" with a marketable commodity in a workplace. This perspective was suggested by Walter Oi (1973), who proposed that an increase in inputs raises not only the quantity of a marketable output but also the probability of an accident or illness.[1] Sickness and injuries are by-products of work, so that an increase in working hours yields both an increase in the output of a marketable commodity and the incidence of ill-health among workers.

Once again, the investigators for the Health of Munition Workers Committee (HMWC) played a seminal role in investigating the impact of working hours on health. They surveyed munition workers to assess their health and recommended actions to improve it. In the surveys of both women and men munitions workers, the investigators determined that long hours

---

1. Boal (2016) adopts this point of view in an empirical investigation of fatalities and the intensity of work among early twentieth-century underground coal miners in West Virginia.

of work rendered the workers vulnerable to illness and injury. As Agnew (1917) wrote, "long hours appear definitely prejudicial to physical well being" (88). Campbell and Wilson (1917, 118) recognized that surveying workers at their place of work understated the damage to health because those most harmed by the work had left the factories and failed to come under their selective review.[2]

The work of some munition workers involved filling the shells with the explosive tri-nitro-toluene, or TNT, contact with which damaged the workers' livers, causing jaundice. Their skin and hair turned yellow, giving rise to the common name for the women workers as "canaries." It was not only a cosmetic ailment; in some instances, death followed.[3]

Rosina Whyatt provides a personal account of working conditions in these munition factories in Burnett (1974, 125–132). In the factory, she worked from 8:00 a.m. to 8:00 p.m., which was "comforting" to her because she "would not have to get up at five or six o-clock as she had done during her domestic service days." In due course, she was diagnosed with TNT poisoning. Her face, neck, and legs began to swell so much that she could not see and had to be led home. When she returned to work, she was told an accident had happened at the factory; a number of girls had been hurt and several killed by an explosion in the department where she had worked. She remarks that she would have been a casualty if she had been working there.

---

2. Campbell and Wilson (1917) examined 1,326 female workers in eleven factories, while Agnew (1917) examined 3,052 men and boys. The men were ages 41 and up, while the boys were 18 years of age or younger.

3. Ineson and Thom (1985, 100) report 111 deaths of male and female workers from toxic jaundice in 1916, 1917, and 1918.

The HMWC identified long hours of work and the associated fatigue as a contributory cause of TNT poisoning. The committee cited Sir James Paget, who wrote "fatigue has a larger share in the promotion or transmission of disease than any other single causal condition you can name."[4] Similarly, in America, writing in 1909 after surveying research that associated long working hours with fatigue and subsequent sickness and mortality, Irving Fisher (1909) concluded that "The fatigue of workmen is largely traceable to their long work day. . . . A typical succession of events is first fatigue, then colds, then tuberculosis, then death. Prevention, to be effective, must begin at the beginning" (45–47).

Is work fatigue a contemporary issue? Ricci et al. (2007) conducted a nationally representative survey of 28,902 US workers between August 2001 and May 2003, asking of individuals, "Did you have low levels of energy, poor sleep, or a feeling of fatigue in the past two weeks?" Of those surveyed, 11,719 (or 41%) replied affirmatively. Information was also collected on health-related lost productive time, which was defined as the sum of weekly hours absent from work for a health-related reason and the weekly hour-equivalent of health-related reduced work performance.[5] Most lost productive time was not attributable to absenteeism but, rather, to reduced performance at work. The incidence of lost productive time attributable to fatigue was greater for full-time workers than for part-time workers.

---

4. James Paget was an eminent nineteenth-century pathologist and surgeon whose work resulted in several diseases bearing his name. The committee's judgment is expressed in HMWC (1919, Section X, Sickness and Ill-health, 61).

5. Health-related reduced work performance was determined by indicators such as the time between arriving at work and starting work on days not feeling well, by the time in lost concentration at work, and by the time not working at work.

A personal description of fatigue from work is supplied by Philip Da Vinci, a lawyer at Legal Aid in Chicago. Da Vinci worked on behalf of the indigent, calling finance companies, trying to defend people who had signed contracts, not knowing what they had signed because they couldn't read English. "You can work four days straight, sixteen hours a day, and never feel tired. Until your eyes start falling out, and then you know you have to go to bed." Da Vinci reports that there is a "burn-out process" such that "guys last here two years and they just burn out. It's just physically too much—and emotionally, God!" (from Terkel 1975, 537–539).

There is now an extensive literature on the association between long working hours and health. The research on this topic is so large that several surveys have been written reviewing what is known. Examples are provided in the review by the Health and Safety Laboratory (2003) in England and the survey conducted by the Centers for Disease Control and Prevention (2004) in America. In view of the Occupational Safety and Health Act of 1970 that defined the mission of the US Occupational Safety and Health Administration (OSHA) as one of ensuring that employees are working in places "free from recognizable hazards that are causing or likely to cause death or serious harm", Anderson (2004) argues that the connection between long work hours and deleterious health outcomes is now so unequivocal that OSHA should investigate and perhaps prosecute employers who require long hours of their employees."

One branch of the research has focused on workers in particular occupations, such as medical workers, police officers, and long-distance truck drivers.[6] For instance, after several years

---

6. On nurses, see Rogers et al. (2004); on other medical workers, see Barger et al. (2005), Czeisier (2005), Landrigan et al. (2004), and Spurgeon and Harrington (1989); on police,

deploring the length of hours worked by hospital physicians, in July 2003, the Accreditation Council on Medical Education (ACGME) in America introduced regulations that had the effect of reducing average weekly hours of interns from 71 to 67 hours (Landrigan et al. 2006). Exploiting the fact that the new regulations affected the hours of work of those in teaching hospitals, but not in the non-teaching hospitals and defining Δ_t as the difference in the mortality rates between teaching hospitals and nonteaching hospitals during period t, Shetty and Bhattacharya (2007) found a drop in Δ_t of 3.75% from the January 2001/July 2003 period to the July 2003/December 2004 period, with a larger reduction in mortality for patients older than eighty years.

Another well-defined group of workers whose health has been tracked over time are British civil servants, the Whitehall I and Whitehall II cohorts. The Whitehall I study followed more than 18,000 men over a decade starting in 1967, all office workers. During this period, mortality rates among those in the lowest employment grades (workers such as doorkeepers and messengers) were three times higher than those in the highest grade (workers such as senior administrators). After controlling for other factors predictive of mortality (such as smoking, physical activity, obesity, stature, and initial health status), the mortality difference between the men in the top and those in the bottom grades narrowed but did not disappear.

A second cohort of British civil servants, Whitehall II, consists of 10,308 workers (two-thirds of whom are men and one-third women) first interviewed in 1985. The longitudinal study is being

---

see Vila (2006) for America and Cooper et al. (1982) for Britain; and on truck drivers, see McCartt et al. (2000)

directed by Michael Marmot of University College London. Following are some of the findings from Whitehall II, an outline of which is found in Marmot and Brunner (2005).

Cognitive Function

A subset of the abovementioned civil servants (consisting of 2,214 employees) were given cognitive tests in 1997–99, with a second wave of tests in 2002–04. The tests covered short-term memory, mathematical reasoning, and inductive reasoning. Virtanen et al. (2009) investigated whether the cognitive scores in 1997–99 were associated with working hours and also whether the changes in the cognitive test scores from 1997–99 to 2002–04 were associated with changes in working hours over this period. In so doing, the investigators took account of the following characteristics of the employees: their gender, age, marital status, occupational grade within the civil service, highest schooling attainment, earnings, smoking, alcohol consumption, physical health status, sleep disturbances, physical activity, and a variable indicating the demands of the job.

Both in 1997–99 and in 2002–04, those usually working more than 55 hours performed significantly worse on two of the five cognitive tests compared with those working 40 or fewer hours. Moreover, those working more than 55 hours recorded a decline in performance in the reasoning test and vocabulary test over the five years. By way of comparison, the difference in cognitive performance of those working long hours compared with those working 40 or fewer hours was about the same in magnitude as between those who smoked and those who did not, smoking having already been determined a relevant factor in cognition decline.

The consequences of long hours on cognitive performance were also examined for 248 production workers in a car plant. Holding constant the employee's age, gender, schooling, alcohol consumption, and job type, Proctor et al. (1996) found that the workers' performance on a short-term memory test and an intelligence test was significantly lower among those who worked more than 8 hours on the day before the testing.

Cardiovascular Disease

Within the general topic of the effect of long working hours on health, considerable attention has been directed toward cardiovascular disease (CVD). A literature providing evidence of a link between working hours and heart disease is at least fifty years old: in 1958, Russek and Zhoman (1958) provided evidence of such an association, shortly followed by Buell and Breslow's (1960) examination of mortality from coronary heart disease in California. The latter found that younger men in nonfarm occupations experienced distinctly higher coronary mortality if they worked more than 48 hours a week. Later, Joseph Eyer (1980) claimed that mortality from heart disease rose in economic expansions and fell in contractions, a relationship he attributed to longer working hours ("overwork") in periods of relative prosperity.

The association between working hours and heart disease in the Whitehall II cohort was investigated by Virtanen et al. (2010). The workers were first questioned about their hours of work "on an average weekday" in 1991–94. Those with any heart disease in or prior to 1991–94 were excluded from the analysis. Over 6,000 full-time workers (ages 39–61 years) were tracked over time: in the baseline period, 54% did not usually work any overtime hours and 10% worked 3 or 4 hours of overtime per day. Those not working overtime worked 7 to 8 hours per day.

The group was surveyed ten years later (in 2002–04) about the occurrence of coronary heart disease (defined as a fatal incident or as a clinically verified nonfatal myocardial infarction or as definite angina). Controlling for age, gender, marital status, occupational grade, alcohol consumption, exercise level, body mass index, sleeping hours, psychological stress, and job demands, those usually working 3 to 4 hours of overtime per day had a (statistically significant) 60% higher probability of coronary heart disease than those not engaging in any overtime work.

Nationally Representative Populations

Although these occupation-specific studies have the advantage of holding constant variables that are characteristic of the occupation and that are difficult to measure and observe, they may present difficulties in generalizing from them. For this reason, the research described next makes use of national representative observations.

An example of a recent study is the research of Sadie Conway and her colleagues. Conway et al. (2016) followed 1,926 American workers in the Panel Study of Income Dynamics over a twenty-five-year period from 1986 to 2011. People were excluded from the analysis if, in 1986 and over the first ten years of positive working hours, they reported cardiovascular disease (angina, coronary heart disease, hypertension, stroke, or heart attack) or if they reported another chronic health condition or disability in 1986.

Measuring weekly hours of work averaged over a period of ten years and controlling for gender, age, schooling, race and ethnicity, marital status, number of children, self-employment, industry, and occupation, Conway found that the incidence of CVD rose with working hours as the hours passed a threshold of 46 per week. Relative to a working week of 45 hours, a week of 55 hours raised the probability of CVD by 16%, a week of 60 hours by 35%, a week of 65 hours by 52%, and a week of 70 hours by 74%. A working week of 75 or more hours doubled the probability of some form of CVD compared with a working week of 45 hours.

Other researchers have also found a link such as this, though the threshold at which the relationship becomes evident may differ from the 46 hours. Using information from seventeen studies across different countries, Kivimaki et al. (2015) followed 528,908 men and women who had no history of stroke at the beginning of the period of their analysis. They controlled for age, gender, indicators of socioeconomic status, smoking, alcohol consumption, body mass index, physical activity, and other indicators of health. They found that, compared with those working 35 to 40 hours a week, those working 41 to 48 hours had a 10% higher probability of a stroke, those working 49 to 54 hours had a (statistically significant) 27% higher probability of a stroke, and those working 55 or more hours had a (statistically significant) 33% higher probability of a stroke.

Using the National Longitudinal Survey of Youth that initially interviewed men and women ages 14 to 22 years in 1979, Dembe and Yao (2016) investigated the long-term effects of long working hours by examining the association between various chronic diseases and hours of work over thirty-two years. By 2010, all individuals were age 40 years or more. Each individual's usual full-time weekly hours of work were averaged over the years. Those who worked part-time (meaning, in this study, fewer than

30 hours a week) were excluded. Of the 7,492 workers included in the analysis, 28% worked an average of 30 to 40 weekly hours (the reference category), 56% worked an average of 41 to 50 hours, 13% worked an average of 51 to 60 hours, and 3% worked an average of more than 60 hours per week.

The relationship between average weekly working hours and eight different physician-diagnosed diseases was examined: coronary heart disease; cancer other than skin cancer; arthritis or rheumatism; diabetes; chronic lung disease; asthma; depression; and high blood pressure or hypertension. In each case, the following regressors were included in addition to working hours: gender, age, race, schooling, family income, years worked in labor markets, smoking, and occupation. No statistically significant link was measured between working hours and four of the doctor-diagnosed diseases: chronic lung disease, asthma, depression, and hypertension. The probability of the other four diseases was significantly higher for those working longer hours: those working more than 50 hours a week had a 68% higher odds ratio of heart disease than those working 30 to 40 hours; the probability of (non-skin) cancer and the probability of arthritis rose monotonically with hours, such that those working between 51 and 60 hours had twice the odds of developing cancer as those working 30 to 40 hours. When the relationship was fitted to women and men separately, women working more than 40 hours a week were substantially more prone to develop disease than men working these hours.[7]

---

[7]. The rising incidence of mortality from cardiovascular disease in Japan attributable to long working hours and accompanying stress has resulted in a new term: *karoshi*, meaning "death from overwork." See Uehata (1991) and Iwasaki et al. (2006).

Injuries and Accidents

The relation between working hours and the incidence of injuries and illnesses of workers in recent years has been studied by Dembe and colleagues (2005), who followed 10,793 workers over time from the National Longitudinal Survey of Youth, tabulating their injuries from 1987 to 2000. Taking account of differences among these workers in age, gender, region, occupation, and industry, those working overtime had a 61% higher hazard rate of injuries relative to those not working overtime.

In 2003, in Korea, legislation was passed reducing the standard working week from 44 to 40 hours. The new regulations were phased in over several years, depending on the industry and size of the workplace. After confirming that working hours did fall, Lee and Lee (2016) found that the reduction in hours was correlated with a fall in accident rates: a cut of one working hour is associated with an 8% lower accident rate, the effect being larger in smaller workplaces.

On the association between working hours and accidents, consider the experience of Bert Coombes (1939), a coal miner in South Wales, born in 1893 and died in 1973:

When he started as a miner before the First World War, his shift lasted eight hours either during the day or at night, and his tools consisted of a pick and shovel. During the First World War, the first cutting machine was introduced and he was selected to operate it. The company was anxious to generate revenues to cover the cost of the machine and, as the demand for coal was strong during the war, his working week was lengthened. He worked a double shift each day, "crawling on our knees under low height for sixteen busy hours" (113)—and the week-end break lasted from two o'clock on the Saturday until six o'clock on Sunday morning. During one week, he was in bed for only eight hours. He dozed at work "very many times" and felt he did not care "if the roof did fall on me, as long as the end of that torture to keep awake came swiftly." Accidents occurred and, during one week, he was "completely buried on three occasions." The greater production that followed the use of the cutting machine and the long working hours induced management to cut the piece rate. The miners protested, but as "they have no money in reserve and are not allowed a quarter, or a half-year, to get their bills paid," the lower piece rates came into effect.

Robert Armstead's (2002) descriptions of his father's working life as a coal miner in Marion County, West Virginia, in the 1920s and 1930s are similar to Bert Coombes's account in Wales. Armstead writes that his father would leave home at 5:00 a.m. to walk to Grant Town Mine, where he would put in a twelve- to sixteen-hour day. "Many times he was too exhausted to eat supper" (34). Roof falls were common, and they maimed and killed. When the roof fell or water seeped into a miner's work area, it took two to four days to clear the debris and drain the water. Then the face had to be secured before mining could start. During this time, no coal was mined and loaded, so the miner's pay (in the form of company scrip) stopped. The overtime premium required by the Fair Labor Standards Act (FLSA) was ignored by small and remote mines, while the larger mines responded with further mechanization.

The Well-Being of the Household

The damaging effects of long hours on a worker's health may have consequences for the household of which he or she is a member. The literature examining this link goes under the name of "worklife balance." The accounts in this literature often describe married couples straining to find time for anything but market work. Indeed, the US Bureau of Labor Statistics reports that, among married couples, the sum of hours of weekly market work by wives and husbands rose from 56 hours on average in 1969 to 67 hours in 2000.[8]

It is not surprising, therefore, that essays and books have appeared in recent years (Boushey 2016, Fried and Hansson 2010, and Schulte 2014 are examples) lamenting the poorly defined boundary between time at market work and time for other activities, and calling for public policies to redress these developments. In this vein, White et al. (2003) used surveys of British employees in 1992 and 2000, in which the employees were asked responses to statements such as "My job allows me to give the time I would like to my partner/family." Five categorical responses were presented, and the answers were combined to form an index of the degree to which an individual's paid employment had undesirable effects on the quality of his or her home life. With 1,474 employees in 1992 and 1,915 employees in 2000, and controlling for a large number of other features of the job and for a number of demographic characteristics, the researchers concluded that for men and for women, in 1992 and in 2000, in accounting for variations in this index, "actual hours worked proved to be by far the strongest explanatory variable in this analysis" (188). The inference was that longer working hours were associated with lower household well-being.

All in all, there is considerable evidence to the effect that longer hours of work have deleterious effects on worker health and quality of life. There is no single level of hours at which these effects are manifest. It is better understood as an increase in the probability of ill-health that rises as hours lengthen.

---

8. Reported as part of the Bureau of Labor Statistics' report "Working in the 21st Century"; see www.bls.gov/opub/working/page17b.htm.

7

The Association Between Working Hours and Hourly Earnings

The Demise of a Basic Issue

Economists have long been interested in the association between hours of work and the hourly earnings of workers. At one time, economists recognized that this association could reflect the preferences and behavior of employers or it could reflect the preferences and behavior of workers. There was an identification problem that required resolution. In recent decades, this issue has largely disappeared from economists' concerns and the association is usually interpreted as one that manifests only the preferences of workers in trading income for work hours. The role of the employer has tended to be downplayed. What was once a critical issue for empirical researchers to address in interpreting observations on workers' hours is now largely neglected. However, the findings of chapter 4 challenge the minor role assigned to the employer in most contemporary research.

The empirical research in chapter 4 was designed to determine how hours of work enter a firm's production function. It was found that, when other things are held constant, the effect of an increase in hours is typically to increase output, but this increase in output is smaller when hours are "long" than when hours are "short". Expressed differently, the marginal product of hours falls as hours lengthen—the law of diminishing returns operates with respect to work hours.

The law of diminishing returns as applied to work hours is important because the net revenue maximizing owner-manager, the pillar of the economic analysis of production in a market economy, will set hours at which the marginal product of hours equals the marginal cost of hours and where the marginal product of hours is declining as hours increase. If the marginal cost of hours consists principally of the hourly earnings paid to each worker, then the purposive owner-manager's choice of hours is where the marginal product of hours equals hourly earnings.[1]

Because the marginal product of hours is larger when hours are "short" than when hours are "long," if the owner-manager is required to pay higher hourly earnings, she will respond by scheduling shorter hours where the marginal product of hours is higher and where it equals the higher hourly earnings. In this way, this employer's demand for hours is closely linked to the marginal product of hours and, when other things are unchanged, longer hours will be scheduled when hourly earnings fall and shorter hours will be scheduled when hourly earnings increase. The demand for hours depends negatively on hourly earnings.

If the owner-manager employs other inputs, then the prices of these other inputs and the price of the output she makes also affect her demand for hours. Suppose q denotes the price of each unit the owner-manager produces and sells, w is the hourly wage paid to each employee, and r is the unit price of other inputs.

---

[1]. For profits to be positive, the marginal product of hours must also be less than the average product of hours for this type of enterprise. In figure 2, hours must exceed $H^\wedge$.

When q, w, and r are all given to the firm, her demand equation for hours per worker is

(29) $$H = f(q,w,r)$$

The owner-manager also has analogous demand functions for the other inputs, including the employment of workers. As far as equation (29) is concerned, the assumption of net revenue maximization, coupled with what has been found about how hours enter the production function, implies that an increase in w reduces this owner-manager's demand for hours of work.[2]

Indeed, this suggests another reason to link the downward movement of hours of work and the upward movement of wages in the nineteenth and early twentieth centuries: as hourly wages rose, owner-managers would be induced to reduce their demand for hours unless their demand functions shifted. The existence of an association between the hours that individuals work and their hourly earnings goes back to the eighteenth century in Britain, at least among workers in the domestic system: "The association of high wages with short hours is documented extensively throughout the eighteenth and into the nineteenth centuries" (Bienefeld 1972, 28).

Nowadays, most workers in America and Britain are employees and most employees are paid according to the time they spend at the workplace. The association between hourly pay and working hours is sometimes derived from observations of workers over calendar time (as in Bienefeld's statement) and sometimes with observations across workers at a given time. However, the interpretation of this association is not straightforward. One interpretation is that it reflects the demand for hours by owner-managers.

---

2. Ehrenberg (1971) estimated equations such as (29) to understand the incidence of overtime hours across US manufacturing industries.

There is another interpretation, though: it maps the preferences of workers for income and "leisure" where by leisure is meant hours not worked in the market. What follows draws, in part, on Pencavel 2016a.

This second interpretation arises from the characterization of the typical consumer-worker as selecting his hours of work, H, and consumption bundle when constrained by a linear income-expenditure identity in which the prices, p, of consumer goods, his hourly earnings, w, and his nonlabor income, y, are predetermined variables. According to this orthodox and unembellished model of labor supply, the consumer-worker does the best he can with what he has, and this yields a supply of hours function as follows:

$$(30) \qquad H = g(p, w, y).$$

Under well-known conditions, the reasoning behind equation (30) implies that an increase in y results in fewer hours supplied to market work, while an increase in w may induce the worker to want to work more hours or fewer hours or not to want to change his hours. This ambiguity in the sign of the effect on H of an increase in w arises because of countervailing substitution and income effects: a higher w raises the opportunity cost of each hour not worked, and this encourages a substitution of work hours for the more costly "leisure" hours; at the same time, a higher w raises the value of the worker's resource (time) endowment, enabling him to work fewer hours for the same labor income, and this inclines the worker to work fewer hours. The *substitution effect* and the *income effect* work in opposite directions when hourly earnings rise: the former predisposes the worker to work more hours and the latter to work fewer hours. Of course, the consumer-worker also has demand equations for the commodities he buys with his income. We neglect this, although a faithful expression of this model has implications for consumer demand, as well as the supply of working hours.

A large literature has been developed that furbishes this spare version of the consumer-worker model and investigates complications such as nonlinear budget constraints, the appropriate treatment of outcomes when hours are zero, household decision-making, the acquisition of skills, and intertemporal issues. There is much to applaud in this work, but little research has been directed in recent years toward distinguishing the preferences of consumer-workers from the preferences of employers. That is, the existence of a demand for hours relation that posits an association between H and w as in equation (29), and a supply of hours relation that also posits an association between H and w as in equation (30), presents a problem of identification: when economists organize observations on the hours of work and hourly pay of workers (controlling perhaps for other variables), are they estimating equation (29) or equation (30), or some amalgam of the two?

The answer turns on what variables are excluded from the estimating equation. If the price of output, q, or the price of an input, r, enters meaningfully into the fitted hours equation, this suggests the relationship is not a supply-of-hours equation because these variables are excluded from equation (30). If nonwage income, y, tends to be negatively correlated with hours, this implies that the hours–earnings relationship is not equation (29), the demand for hours function, because the owner-manager's demand equation excludes y.

Note that the estimates of equation (23) in table 4.19 provides a classic example of what may be expected of a demand-for-hours function: the nonwage income of employees is excluded, as is the price of consumer goods; instead, the equation includes hourly earnings, the price of the firm's output, and the price of another input (logs). In fact, these estimates exhibit the zero homogeneity property of input demand functions without imposing this restriction on the estimating equation. The negative sign on the logarithm of hourly earnings is also consistent with a supply-of-hours equation in which the income effect outweighs the substitution effect, but the researcher needs some imaginative reasoning to reconcile with a supply equation of hours of work the positive effect on hours of higher output (plywood) prices and the negative effect on hours of higher prices of the input of logs.

With these remarks as prologue, consider some of the empirical research in the last fifty years on hours of work and hourly earnings.

Marvin Kosters

In terms of its attention to economic principles, modern research on this class of issues begins with Marvin Kosters' PhD thesis at the University of Chicago, subsequently published as a Rand Report (1966). Essentially, Kosters took Gregg Lewis's explanation for the time-series movements in hours of work and real wages (described in chapter 2 and figure 1) and converted it to an explanation for cross-sectional differences in hours and wages across individuals.[3]

From the 1960 US Census of Population, Kosters selected observations on husbands ages 50 to 64 years who worked in 1959 and in the week preceding the Census. He analyzed both hours worked per week and hours worked per year, the latter being the product of weekly hours and weeks worked in 1959.

---

3. Lewis was a member of Kosters's dissertation committee, and in the acknowledgments, Kosters thanks Lewis for "important aid in clarifying the theoretical framework."

The Association Between Working Hours and Hourly Earnings

Insightfully, he recognized that, in principle, hours of work ought to be "standardized for intensity of effort" (22). Hourly earnings were measured by dividing annual earnings by hours worked per year. "Income of the family from nonemployment sources" (23) served as the measure of y. Kosters was explicit about his presumption that these measures of hours worked of the men "approximate their desired labor supply" (25).

He was candid in writing that he presented only those results of estimating equations that did not contradict his prior beliefs: "only those regressions for which the sign of the estimated coefficient on y was non-positive was reported" (27). Or, again, "coefficients [on y] of sufficient magnitude with the appropriate sign were in general not obtained" (35). Thus, Kosters found that higher nonlabor income is not associated with shorter hours among these workers. If higher y does not reduce hours of work, then the satisfaction of the consumer-worker's budget constraint requires that non-working time is inferior and a dollar increase in nonlabor income increases the consumption of commodities by more than one dollar. This might prompt the question, "Is this a demand for hours equation that is being estimated?"

Indeed, in most instances, Kosters found that the estimated effect of higher wages on working hours is negative, a finding fully compatible with a negatively sloped demand-for-hours function.[4] Whereas Lewis had addressed the identification problem, Kosters does not face the problem squarely and overlooks the possibility that he is really fitting a demand-for-hours equation, not a supply-of-hours equation.

---

4. If this is a supply-of-hours equation as Kosters and others presume, if $\partial H/\partial y \geq 0$ and if $\partial H/\partial w < 0$ as Kosters finds, then the implied substitution effect is nonpositive, a flagrant refutation of the theory's implication. See Pencavel 1986.

As an illustration of his results, here is the least-squares equation where H denotes the weekly hours of 8,467 husbands:

$$log(H) = \beta Z - 0.043 log(w) - 0.010 log(y)$$
$$(0.009) (0.011)$$

where Z stands for nine other regressors in the equation (including demographic indicators, industry dummy variables, and regional dummy variables) and β represents their estimated coefficients. Estimated standard errors are in parentheses beneath their estimated coefficients.[5] Is the appropriate inference from these estimates that the uncompensated wage elasticity of the supply of these workers' hours is -0.043, or is this the wage elasticity of employers' demand for hours? The latter is consistent with our inability to reject the null hypothesis that differences in nonlabor income are uncorrelated with hours.

In the sixty years or so since Kosters's dissertation, a large part of the literature—"one of the most extensive in economics" (Keane 2011, 961)—has followed his procedures and replicated his findings. That is, differences among workers in their hours of work are assumed to reflect only their budget-constrained hours-income preferences, and in a number of cases, his result of a nonnegative association between their nonlabor income and working hours has been reproduced. Frequently, researchers have sustained Kosters's result that the partial correlation between working hours and hourly earnings for men is negative.

Of course, some have found a positive association for working men between hours and earnings, but this tends to be a fragile result. For instance, when MaCurdy (1981) adds calendar-year

---

5. This is equation (10) in table 4 of Kosters (1966, 30).

dummy variables to his hours-wage (in first differences) regressions, the point estimate on wage changes becomes smaller than its standard error. In the context of his model, this implies the hypothesis of no intertemporal substitution in hours cannot be rejected by conventional criteria.

Calendar-year dummy variables also present problems for Browning et al.'s (1985) lifecycle estimates. At least since Bry (1959), movements in the demand for hours over the business cycle have been recognized, and these changes would account for the relevance of calendar-year effects in equations accounting for changes in hours of work over time.

Certainly, some economists have been troubled by the assumption that differences in hours entirely reflect workers' constrained preferences. This unease is consistent with the responses of many workers who, when asked, express a preference for working more or fewer hours at their current wage rate (see Stewart and Swaffield 1997, Golden and Gebreselassie 2007, and Bryan 2007). In some jobs, overtime hours are routinely required by the employer (Golden and Wiens-Tuers 2008). This has induced some researchers to introduce into equations of working hours constraints by employers on working-hour choices made by workers. This class of work might find it helpful to introduce the price variables in equation (29) to account for differences in hours.

There are a few exceptions to the generalization that the identification problem in hours-wage research has been neglected, and here are some notable ones.

H. Gregg Lewis, Again

Even though Lewis's identification assumptions in his 1957 work were tacitly accepted by much of the economics profession, Lewis himself retracted his earlier belief that the interests of employers play no part in the determination of workers' hours of work. In a Spanish-language journal in 1969, he published an article that presented a model in which the preferences of employers affect the setting of working hours in labor markets. An English-language version of this paper with the title "Employer Interests in Employee Hours of Work" has circulated for over forty years. In this article, the owner-manager is assumed to be willing to offer higher wages to a worker for longer hours so as to economize on fixed costs of employment. For his part, a worker has to determine which bundle of earnings and hours is most desirable. A single employer and a single worker take the market wage-hours function as given.

This market-wage function (what Lewis called the "market equalizing wage curve") reflects the underlying preferences of workers and employers, but it is neither the typical worker's supply-of-hours equation nor an owner-manager's demand-for-hours equation. It is a joint envelope of a set of workers' value functions and another set of employers' offer functions. This model was reconfigured by Sherwin Rosen (1974) to help understand the fact that consumer goods (and jobs) have characteristics and attributes that are linked to their prices (and wages). Rosen's hedonic price function, in which the price of a commodity varies with a scalar index of the characteristic, was Lewis's market-equalizing wage curve, as Rosen acknowledged. Though Lewis's second model of the labor market was used extensively in research on hedonics, it had little impact on empirical research on hours of work. Usually without reference or recognition, economists adopted Lewis's 1957 framework and imitated Kosters's application to cross-sectional observations. Questions of identification were simply not addressed or even recognized.

Martin Feldstein

One recognition of the identification problem was the attempt by Martin Feldstein (1968) to estimate a supply-of-hours function using British observations in 1965. He explicitly outlines the identification issue and suggests a solution to it: use observations from a single market where an unchanging supply-of-hours function is intersected by shifting demand functions for hours. He implemented this in defining a labor market by an occupation in a specific industry that is located in a particular region in England. There were eleven such groupings of observations and thus eleven least-squares estimates of the hours–earnings slope. Eight of these slope estimates were negative, though most were of doubtful statistical significance. Feldstein concluded "the identification problem persists" (78).

Sherwin Rosen

Another clear recognition of the identification problem in associating hours of work and wages was Sherwin Rosen's (1969) empirical study of hours and hourly earnings across US industries in 1960. Rosen used data on individual workers from the 1 in 1,000 sample of the 1960 US Census of Population aggregated to allow calculation of industry average values. He specified equations designed to represent the supply of working hours per worker, and another the demand for hours per worker, and he estimated these jointly. He found "demand seems to fit the data better than supply" (269) with estimated wage elasticities of the demand for hours (evaluated at mean values) of between -0.35 and -0.80. "Demand is more elastic than supply and the assumption of infinite demand elasticity [Lewis's 1957 identification assumption] is untenable in industrial cross-sections," (269) he wrote. Once again, the estimated coefficient attached to the variable designed to capture the negative effect of nonlabor income was "approximately zero."

Michael Abbott and Orley Ashenfelter

Michael Abbott and Orley Ashenfelter called upon Lewis's (1957) first model to justify the interpretation of the hours–wage equation as a supply function when fitted jointly with a system of commodity-demand equations. Using aggregated data over the years from 1929 to 1967, they tend to estimate negative (uncompensated) wage elasticities, a result consistent with Lewis's (1957) interpretation. In this instance, consumer goods' prices are entered as determinants of hours of work, and at least movements in some of these prices seem to be associated with movements in hours of work. Because such prices are excluded from a demand-for-hours equation, this association strengthens the case for this hours-of-work equation being what it is purported to be—a labor-supply function.[6]

Dora Costa

Dora Costa (2000a) also recognized the identification problem and questioned the orthodoxy that the hours–wage mapping traces out the supply curve of a typical worker's hours. From published American state surveys in the 1890s of nonfarm wage earners, she extracts observations on workers' wages and their usual daily hours of work, their age, and their gender. At the time

---

6. To be precise, consumer prices, designated by p in equation (30), are excluded from the demand-for-hours equation (29), in which the price of the firm's output, q, is the relevant argument.

The Association Between Working Hours and Hourly Earnings

of these surveys, none of the states in her data had passed legislation regulating working hours. Separating men from women, she places these observations into bins defined by their decile in the hourly wage distribution and, within each wage decile, she averages their daily hours of work. She does the same for workers drawn from a supplement to the 1991 US Current Population Survey. The resulting hours–wage patterns for the 1890s and for 1991 are shown in figure 21 for men and in figure 22 for women.[7]

In the 1890s, for both men and women, the hours–wage association is strongly negative: low-wage workers worked the longest hours. In 1991, for women, the association is mildly positive. While low-wage women workers were working over 3 hours a day less in 1991 than a century earlier, high-wage women are working almost the same hours (8.72 hours) as they were in the 1890s (8.95 hours). For men, in 1991, there is little difference in daily hours between the workers with wages in the ninth decile (8.64 hours) and those in the fourth decile (8.61 hours). In 1991, low-wage men are working almost 3 hours less per day than low-wage men in the 1890s.

Costa doubts that the appropriate interpretation of these relationships is as workers' supply of hours, and she remarks "the data needed to identify a labor supply response of the 1890s are unavailable" (177). She reports on equations fitted to her 1890s observations, in which hours are regressed on hourly wages, and on indicators of nonlabor income. She estimates "income elasticities that are close to zero" (176) and recognizes that "it is not clear whether I am identifying a labor supply or labor demand curve."

---

7. These figures are for all working men ages 25 to 64 and for all working women ages 18 to 64. She weights the observations in the 1890s so that their occupational distribution is the same as that in 1991.

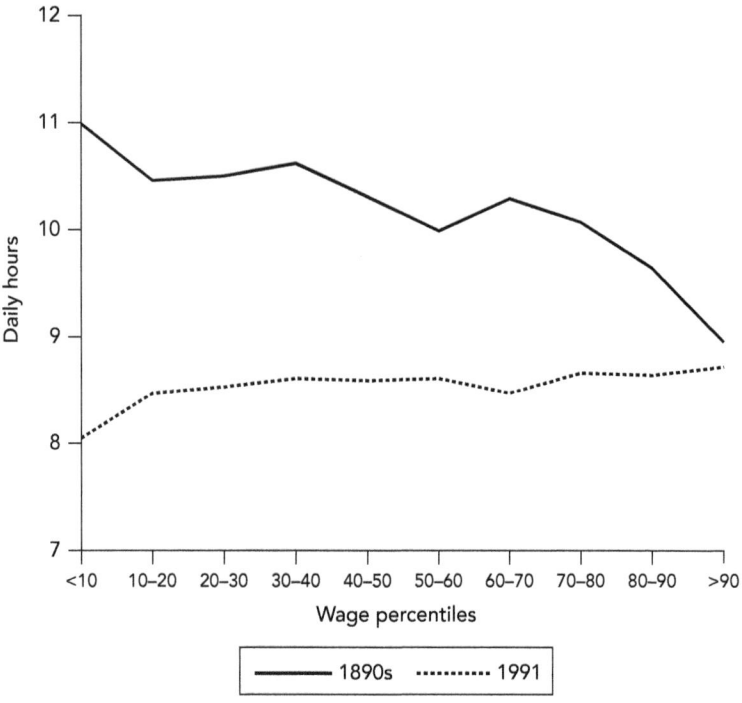

FIGURE 21 Usual Length of Work Day by Hourly Wage Percentiles: Men aged 25–64, 1890s and 1991.

From: Dora Costa (2000a).

The Textbooks

These scholars—Martin Feldstein, Sherwin Rosen, Michael Abbott and Orley Ashenfelter, and Dora Costa—recognized the issue of identification in hours–earnings regressions, but most other researchers did not acknowledge its existence. Over the last forty years or so, the dominant line of scholarship has taken for granted that, fitted to a cross-section or a panel of workers,

The Association Between Working Hours and Hourly Earnings

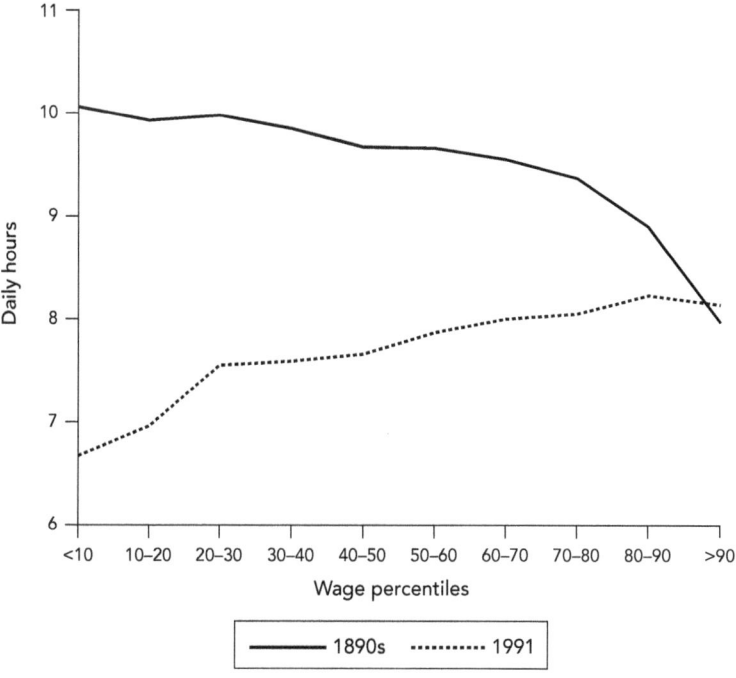

FIGURE 22 Usual Length of Work Day by Hourly Wage Percentiles: Women aged 18–64, 1890s and 1991.

From: Dora Costa (2000a).

a regression of their hours on their hourly earnings maps the constrained preferences of workers.[8]

Indeed, it is remarkable that many now use the words "labor supply" as a synonym for "hours." Contemporary labor economics textbooks display this practice so that students, in turn,

---

8. Keane's (2011) authoritative survey of this research is an exception. Killingsworth (1983, 437) recognized the risk of ignoring "potentially important secular trends in demand."

adopt the habit without being instructed on the tacit assumption being made.

It is curious because, elsewhere in the text, these books discuss the fact that hours of work tend to rise in expansions and fall in contractions, and they attribute this to employers' actions in adjusting inputs to meet the fluctuating demands of their customers: movements in the demand for working hours are drawn upon to explain variations in hours over the business cycle, but differences in the demand for hours play no role in accounting for differences in hours in cross-sectional observations. It is odd because some workers are employed in industries that are very susceptible to the business cycle while other workers are employed in industries more protected from such influences. Hence, one might expect a consideration of the demand for hours in understanding the pattern of cross-sectional variations in hours worked.

The transformation in economists' thinking about hours of work is evident when contrasting the pre-1960s texts with contemporary textbooks. Whereas the texts of today treat the words "working hours" and "labor supply" as interchangeable, the older texts take it as a given that the employer occupies an important role in the determination of working hours, and they attend to the manner in which hours enter the production function. For instance, Reynolds's text (1954, 254) draws a production function (output as a function of weekly hours of work) that closely resembles those estimated in this book.

Another older textbook (Lester 1946) opens the chapter on "Hours of Work" with the statement, "The length of the work day and the work week are of importance to workers and to society. Long hours of toil may affect the worker's health and reduce the length of his working life. Such hours may also stunt his growth as a citizen by failing to permit him sufficient time for social, cultural, and political activities." This recognition of the larger role that working hours play in the lives of workers is missing from texts nowadays. Lester adds "Reduced hours decrease the absence from work because of sickness, the number of accidents per hour, and the percentage of defective output. Increases in normal working hours have the reverse effects" (351).

In this climate of thought, it is not surprising that, when Edward Denison (1962) undertook his well-known investigation into the sources of American economic growth, he hypothesized that, at 49 weekly hours of work—approximately the hours typical of American workers in 1929—a reduction in hours would be fully offset in work effort as to leave output unchanged. When weekly hours were about 40 in 1957, Denison guessed that a 10% reduction in hours would lead to a 6% reduction in output—approximately the value that appears in estimating Cobb and Douglas's augmented production function to Vernon's observations described in chapter 4 (see figure 7). According to Denison, the contribution of reductions in working hours to the growth of real national income per worker-hour between 1929 and 1957 was more than twice that of the increase in physical capital.

Denison was aware of the fragility of his guesses, and he offered the judgment that "Few studies offer more promise of adding to welfare and contributing to wise decisions in a matter that may greatly affect the future growth rate than a really thorough investigation of the present relationship between hours and output. Such an investigation would deal with a wide variety of occupations and industries operating under different conditions" (39).

In Britain, the analogous research (Matthews et al. 1982) also acknowledged the productivity benefits of shorter hours. The authors postulated that, at the aggregate level, there was no loss in the effective labor input from reductions in hours before 1914. Subsequently, as hours fell further, the productivity benefits of shorter hours diminished until they were zero after the Second World War. The authors describe these estimates as "largely guesswork" (96).

If the issue of distinguishing between the supply of work hours and the demand for work hours merited research attention in the 1960s and 1970s, why did it virtually disappear during the next forty years? The nature of the research became one of measuring the parameters of a presumed relationship, not of testing hypotheses that the relationship described one structural equation and not the other. It was as if there were an agreement in the profession not to raise these awkward identification questions.

Perhaps the most palpable feature of the frequency distribution of working hours is the spike in hours at which overtime wage premia apply. In America, for covered workers, currently this is at 40 hours per week. What configuration of the determinants of workers' income–hours preferences leads to almost half of American workers choosing 40 hours year after year? (see table 1.1). In other words, how can a model of market work behavior that chooses to ignore the preferences of employers account for this most visible of characteristics of working hours?

This is demonstrated by van Soest et al. (1990) using observations on Dutch workers in 1985. They show that the implications of a purported labor-supply equation that neglects employers' preferences concerning working hours results in a poor correspondence with observed hours. This induces the authors to introduce restrictions on the set of hours from which workers may choose. Not surprisingly, when certain hours are prohibited from the choice set, a better correspondence between the implied hours and observed hours follows. Although the authors call these "demand-side restrictions," there is little in the way of economics to justify the restrictions, such as those arising from differences in the right-hand-side variables in equation (20).

On the presumption that the hours–earnings association maps consumer-workers' preferences, considerable effort has been directed toward accounting for taxes and transfers in the measurement of the net returns facing workers. Such considerations apply to the demand function for hours, equation (29), insofar as payroll and other taxes are levied on firms. Yet this has attracted far less attention from researchers than the analogous issue for consumer-workers.

There is much of value in this research on hours and wages that presumes to reflect the preferences of workers. However, the research would be enhanced if some effort were directed toward confirming that the fitted equations describe one set of preferences or the other. If a regression of hours of work on wages and nonwage income yields a positive or zero coefficient on nonwage income, the researcher needs to question the interpretation of the relationship as a labor-supply equation. If, moreover, the hours of work of individuals appear to be associated with the prices of inputs and output at their workplaces, the case strengthens for this hours–wage relationship as a demand-for-labor equation. It would be an advance if a fitted hours–wage relation were not presented as a labor-supply function as a matter of faith, but if its interpretation as a reflection of individuals' income–work preferences were demonstrated.

The Typical Estimated Relation—A Hybrid Equation?

So what has been estimated when researchers present estimates of an hours–earnings equation? I conjecture that, in many cases, the fitted equation is neither a supply-of-hours equation embodying the constrained work–income preferences of workers nor a demand-for-hours equation reflecting the owner-managers' objectives and the manner in which working hours enter the production function of workplaces, but rather a mélange of the two. Here is a simple expression of this conjecture in which the supply and demand equations are assumed to be conveniently linear in the logarithms of the variables:

Suppose H_i denotes the hours worked of worker i and w_i represents worker i's hourly earnings, and suppose the demand-for-hours equation of many workers may be described by the following equation:

$$log H_i = \alpha_0 + \alpha_1 log w_i + \alpha_2 X_i^D + \varepsilon_i^D$$

where $\alpha_1 < 0$ and X_i^D stands for predetermined variables such as the prices of output and inputs and firm- or industry-specific effects that shift the demand-for-hours function. The effects of the production technology on hours worked are also embedded in the equation.

Analogously, let the supply-of-hours equation for these workers be

$$log H_i = \beta_0 + \beta_1 log w_i + \beta_2 X_i^S + \varepsilon_i^S$$

where β_1 may be positive or negative depending on the relative magnitudes of the substitution and income effects of these workers. X_i^S are predetermined variables such as nonlabor income that shift the supply-of-hours function in addition to variables believed to be associated with differences across workers in preferences. Omitted stochastic elements are represented by ε_i^D and ε_i^S.

Suppose λ is the fraction of these workers whose hours conform to the demand function, leaving the remaining fraction $1 - \lambda$ of workers who work the hours that satisfy their constrained preferences. Multiplying the demand-for-hours equation by λ, multiplying the supply-of-hours equation by $1 - \lambda$, and adding the resulting two equations yields

(31) $\qquad \log H_i = \gamma_0 + \delta \log w_i + \gamma_1 X_i^D + \gamma_2 X_i^S + u_i$

where $\delta = \lambda \alpha_1 + (1 - \lambda)\beta_1$ and u_i incorporates both ε_i^D and ε_i^S. δ may be positive or negative.

If it is interpreted as the elasticity of hours with respect to wages in a supply-of-hours function, then δ underestimates the key parameter of employees' preferences, β_1, unless λ is zero. Unfortunately, the value of λ is unknown, though it would be daring to maintain it is zero.

If scholars are fitting equation (31), then differences and variations in δ may be related not to differences in the work-earnings preferences of employees (that is, β_1), as many researchers assume, but to differences and variations in λ. To pursue this possibility, perhaps the most persistent result in this literature is that δ is positive for women and that δ is less, perhaps negative, for men. This could mean that β_1 differs between women and men, or it could mean that preferences are the same but λ is smaller for women than for men. It is difficult to know if this is the case without some knowledge or guesses about the magnitudes of λ for men and women.

Perhaps workplaces where workers are highly complementary for one another, and where starting and ending times for work are the same for all employees, might indicate places where employers unilaterally specify the hours their employees work and where the hours-wage observations lie on the employers' demand-for-hours curves. An assembly line in a manufacturing plant may typify this sort of workplace.

There are noticeable employment differences between men and women in the industries and occupations in which they work, and these may well be related to λ. In America, women constitute 97% of the employment of preschool and kindergarten teachers and 9% of manufacturing industry employees. Is the scheduling of hours of work of preschool teachers more responsive to the wishes of its employees than that of manufacturing workers? In other words, is the share of women working in places where the employees have greater latitude in selecting their work schedule greater than that of men? If so, the positive values of δ often estimated for women workers imply that a greater fraction of women work in activities in which their work preferences are accommodated by their employers and λ is low.

The partial correlation between working hours and hourly earnings among women workers appears to have declined in recent decades. See Blau and Kahn (2007), Heim (2007) and Bishop et al. (2009). From this, should one infer that the work–income preferences are changing or that an increasing fraction of women are working in markets where λ is lower? Once the possibility that a fitted hours–wage relation represents the preferences of employers is permitted, the choice of control variables in the hours equation can be reexamined. The usual practice has been to draw upon large data sets on individual workers and allow for gender, age, and marital status differences in the hours–wage relation, but little more. When a demand function for hours becomes a possibility, the common procedure of neglecting aspects of the workplace (such as the presence or absence of labor unions), the production technology, and the objectives of the organization becomes untenable.

Information Provided by Self-Employed Workers

The identification problem causes the hours–income preferences of employees to be entangled with those of employers. When the goal is to determine the preferences of the workers, this farrago would not arise if the workers were their own bosses. This reasoning suggests that the income–hours preferences of workers might be ascertained from observations of self-employed workers who are able to set their own work schedules within the confines of their customers' willingness to pay.[9] For the self-employed worker, the role of the employer is assumed by the final customer. Here is an illustration of how the analysis might proceed.

Consider an organization managed and owned by those who work in it. Such an organization is a worker co-op or a producer co-op. The co-ops studied here are the plywood co-ops of the Pacific Northwest that were included in the analysis of the plywood production function in chapter 4. In each co-op, all workers work the same number of hours and earn the same hourly returns. For many years, the plywood co-ops in the Pacific Northwest constituted the most durable worker-owned and worker-managed sector in US manufacturing industry. In the 1950s, almost 100% of US softwood plywood was produced in the Pacific Northwest and between one-fourth and one-fifth of that was made in the co-ops (Berman 1967, 93). Since then, the depletion of the old-growth timber forests in the region, the restrictions on logging, and the ensuing rising cost of logs caused the center of US plywood production to move to the American South.

The observations examined here follow plywood mills over a period when the industry was of waning importance. As the industry contracted in the Pacific Northwest, a number of co-ops (and conventional mills) closed. The research uses 55 mill-year observations on 11 co-op mills in even-numbered years between

---

9. As Berman (1967, 228) writes of the plywood co-ops analyzed here, "As owners, the workers have the power to set hours of work to equate for themselves, at least on some sort of average basis, the marginal utility of money income and of leisure."

1968 and 1986 in Washington State.[10] All co-op members were workers in the enterprise and most workers (often, all workers) were members. Each member had one vote in the co-op's decisions in selecting the directors from among the workers. The principal constraints on each plywood mill were the price of its major input, logs, and the price of its output, plywood, both beyond the control of any single mill. Most logs came from federal and state forests, and were sold at public auctions.

Log and plywood prices are volatile because the demand for plywood fluctuates with home and office construction, which is highly cyclical. The mills used different varieties of wood, so that log and plywood prices varied across mills, but much of the variation in these prices is over time.[11] Consequently, the wage paid out of net receipts to the co-op workers also varies with movements in the prices of logs and the prices of plywood. Illustrative of this is the following estimated regression equation, which uses the 55 mill-year observations on the co-ops to relate the logarithm of the real hourly wage of co-op workers, $ln(w_{it})$, to the logarithm of real log prices, $ln(r_{it})$, and the logarithm of real plywood prices, $ln(p_{it})$, controlling for fixed mill effects, v_i:

$$ln(w_{it}) = v_i + \underset{(0.152)}{0.942 ln(p_{it})} - \underset{(0.154)}{0.252 ln(r_{it})}$$

where heteroskedastic-consistent standard errors are in parentheses. The R^2 statistic is 0.577. A mill is indexed by i and a year by t.

---

10. There are more observations on the co-op mills for our analysis of the association between hours and hourly earnings than there were for the production function investigation in chapter 4. This is because some co-op mills lacked observations on the input of raw material logs that were needed to estimate the production functions but are not needed in this section.

11. Thus, for these observations, 93.6% of the variation in the logarithm of log prices and 40.9% of the variation in the logarithm of plywood prices are removed by fixed year effects alone.

These plywood co-ops respond to shocks in input (log) prices and to output (plywood) prices by adjusting their wages, whereas the corresponding adjustments to variations in log and plywood prices in the capitalist plywood mills tend to take the form of changing input and output quantities, including employment and hours.[12] In a number of respects, the workers within each plywood co-op were homogeneous and, furthermore, procedures were adopted that cultivated such consonance: the workers were all men and jobs were rotated. This smoothed administration of the plant, and as the member-workers tended to play an active part in making important decisions, they were predisposed to execute these decisions effectively.

Descriptive statistics on annual hours of work and real hourly wages (in 1967 dollars) on the co-op plywood mills[13] and, to provide a comparison, on the conventional unionized mills in Washington state are provided in table 7.1. There are 55 mill-year observations on 11 co-op mills and 102 observations on 19 conventional unionized mills. At all points of the distribution of working hours in table 7.1, the value for working hours in the co-ops exceeds the value for hours in the conventional unionized mills.[14] As is well known (Carrington et al.1996), the self-employed tend to work longer hours than employees. This is also evident in the 2000 US Census observations in table 1.2.

In the case of the co-op workers, do they work longer hours because they shape their work environment, which makes them less averse to work than employees whose control over their

---

12. See Craig and Pencavel (1992) and Pencavel and Craig (1994).

13. The observations on hourly wages are taken from the records of the mills and represent average values over the year.

14. Moreover, hours in the co-ops are right-skewed while those in the conventional mills left-skewed.

TABLE 7.1 Descriptive Statistics on Annual Hours of Work and Real Hourly Wages in Worker Co-ops and in Conventional Unionized Mills

	Coop mills		Conventional unionized mills	
	Annual hours	Real wages	Annual hours	Real wages
minimum	1.232	1.628	0.864	1.030
25th percentile	1.960	2.947	1.758	3.691
median	2.048	3.299	1.912	4.132
mean	2.086	3.361	1.860	4.364
75th percentile	2.265	3.781	2.000	4.758
maximum	2.920	5.371	2.840	10.031
coefficient of variation	0.145	0.209	0.158	0.269

Note: Hours are expressed in thousands of annual hours and wages in 1967 dollars.

workplace is more circumscribed? Or is the difference between the hours of workers in co-ops and the hours of workers in capitalist mills not a consequence of their greater control over their workplace but, rather, a reflection of self-selection: those workers who are inherently less work-shy are attracted to a cooperative enterprise?[15] Berman (1967, 229) believes that "worker-owned companies tend to attract men who wish to work steadily and the differences in motivation and conditions of work in a cooperative may importantly affect estimation of the disutility of labor."

To describe the relation in these worker co-ops between annual hours of work, H, and real hourly wages, w, a quadratic-in-wages

15. The observations of co-op plywood mills are not distributed across years in the same way as the observations of conventional mills. Therefore, it may be important to control for this difference in computing the difference across mills in annual working hours. When this is done, workers in co-op mills work 33% more hours than workers in the conventional mills.

spline was estimated to the 55 mill-year observations allowing for fixed mill effects, and the implied hours–wages relation (over the range of hourly wages observed) is pictured in figure 23.[16] The hours–wage relationship is negatively sloped at relatively high and relatively low wages, but at between 2,000 and 2,100 hours (roughly 40 to 42 weekly hours), the relationship is close to vertical. In other words, at around 40 hours of work a week, differences in wages are associated with little difference in working hours.

Most of the elasticities of hours with respect to hourly wages implied by the estimates of the quadratic spline pictured in figure 23 are between zero and minus unity. At between 2,000 and 2,100 annual hours, they are close to zero. The 95% confidence intervals around the fitted relationship are given by the dashes in figure 23 and it is evident that, over the range of real wages observed, a monotonic positive relation cannot be drawn within these intervals and a negative or vertical hours–wage slope dominates. Even when negative, the reduction in hours when hourly wages rise is not sufficiently large as to cause annual earnings to fall.

These estimates are compatible with Berman's (1967) close study of these worker co-ops:

> The long weekly hours commonly worked in the co-op plants suggest that in the worker-owned companies the supposed "disutility of labor" plays a negligible role in decisions about hours of work.... [T]he short-run supply curve for hours of labor for shareholder workers might take the shape of a vertical line at the forty-hour-weekly point (showing shareholders willing to work forty hours over a wide range of wage rates), positively inclined at the

---

16. The derivation of the estimating equation and further detail are contained in Pencavel (2015a).

top at some point of high wages ... and negatively inclined at the bottom at some point of low wages. The vertical segment may be an area extending from forty to forty-eight hours. (229)

Findings on some other self-employed workers are similar to those of these plywood co-ops. Thus, in their research on New York taxi drivers, Ashenfelter et al. (2010) estimate the uncompensated elasticity of hours with respect to wages to be approximately -0.23 with a relatively small estimated standard error. Boulier's (1979) investigation of nonsalaried dentists found annual working hours of these dentists to be negatively related to

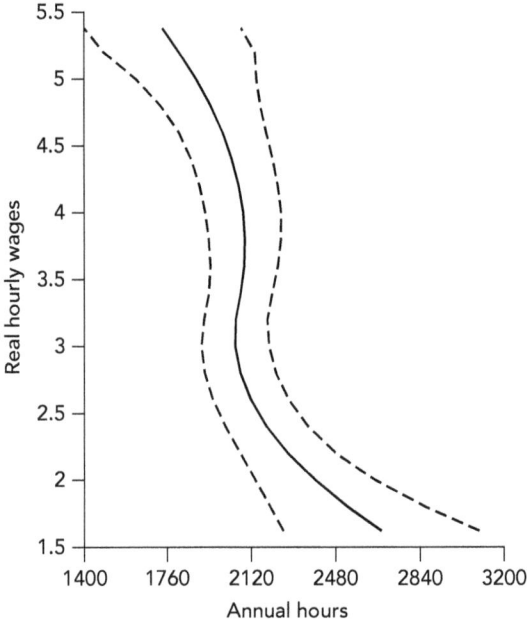

FIGURE 23 Worker Co-ops' Hours–Wage Relation (and 95% Confidence Intervals) Implied by the Estimated Quadratic-in-Wages Spline.

Note: The dashes denote 95% confidence intervals around the fitted spline.

the price of a tooth extraction, the elasticity being -0.049. With 1,456 observations on 286 self-employed men from the Panel Study of Income Dynamics, Parker et al. (2005) report a negative elasticity of annual hours of work with respect to hourly wages of -0.234, though this becomes imprecise when the standard deviation of past wages is introduced into the fitted equation. Is an empirical regularity emerging—that self-employed workers tend to work slightly shorter hours when their hourly earnings increase, although the hypothesis of no change in hours when hourly earnings rise is not greatly at variance with the evidence? Recall that this was also Bienefeld's description (1972, 28) of those working under the domestic system in eighteenth-century Britain.

During the period of analysis of these plywood mills, two mills converted from one type of ownership to another (in one case from a co-op to a conventional mill and in the other from a conventional mill to a co-op). This allows for the examination of the change in hours of work *in the same plant* when the identity of the owners switches between a conventional firm in which ownership and management is controlled by those who supply the financial capital and a cooperative organization in which ownership and management are in the hands of those who work in the plant. In these ownership conversions, working hours are lower when the mill takes the form of a conventional organization in which the workers are neither the managers nor the owners.

When these mills were conventional capitalist workplaces, the employees were covered by a multi-employer collective bargaining agreement, and the hours of work were those chosen by management in response to the wage specified in the union–management contract. To the capitalist management, the hourly wage is a cost and the union's impact on wages and on the owners' net revenues can be mitigated by cutting hours of work. To the worker co-op, the hourly wage is a key factor in a worker's income and the effects of a higher hourly wage on his income can be enhanced by working longer hours (the substitution effect). The wage thus plays a different role in the objectives of the two ownership forms.

Typically, the interpretation of the hours–wage relation requires attention to the issue of identification: is this a demand function or a supply function, or neither? In the case of the plywood co-operatives, with observations at the workplace level, the identification problem is addressed by specifying the decision-makers: the self-employed member-workers in the co-op mill and the capital owner-managers in the capitalist mill. If these and other results using the self-employed may be generalized, workers' hours of work choices change little when their hourly earnings rise or fall.

A Third Class of Explanations

The issue of the appropriate interpretation of an association between working hours and hourly earnings has been presented here as a classic identification problem: do the observations on hours and pay trace a demand function for hours or a supply function of hours, or some mixture of the two? There is another interpretation in which the labor market is characterized as one in which workers and employers bargain over working hours and wages jointly. The bargaining may be explicit when a trade union represents the workers or it may be tacit when workers embody skills that would be costly for the employer to replace.[17]

---

17. A taxonomy of bargaining models involving working hours is found in Earle and Pencavel (1990).

The Association Between Working Hours and Hourly Earnings

Once this possibility is recognized, the decline in working hours and rise in hourly earnings in the century from 1840 in the United States and Britain may be neither movements along an approximately constant demand function for hours nor movements along an approximately constant supply function of hours, but the reflection of the growing yet fluctuating relative bargaining power of workers. Reductions in hours have tended to affect a large number of workers at about the same time, implying, perhaps, economy-wide changes in workers' bargaining power. The pattern of the decline in hours—that is, "hours manifest a tendency to shift from one plateau to another and ordinarily each new standard prevails for a period of years" (Millis and Montgomery 1938, 365)—might suggest periodic changes in the relative bargaining power of labor unions that affect many types of workers.

The working hours of a large class of workers have been on this plateau during recent decades: according to the series in Huberman and Minns (2007), the average weekly hours of British production workers were 42 in both 1970 and 2000 (for men) and the average weekly hours of American workers were 38.8 in 1970 and 43.3 in 2000 (for men) Many of these workers may well have been unionized today if the economy of 2000 resembled the economy of 1970. In other words, it may be no coincidence that, over a period when the US and British economies have grown considerably, the small changes in working hours and small increases in earnings for many workers have come at a time when the scope and role of trade unions in these economies have shrunk considerably. In the determination of hours and wages, trade unions in Britain and America once served to provide focal points around which employers, workers, and legislators organized their thinking. Their role in providing these norms has been diminished by the growth in the extent of unfettered labor markets.

8

Concluding Notes

In accounting for the decline in working hours in both America and Britain dating back to the mid-nineteenth century, among the competing explanations, in this book, pride of place has been accorded to the activities of trade unions—not only in workplace bargaining with employers but also in expressing the preferences of all workers whether unionized or not. The evidence for this has at least three components.

First, in cross-sectional observations, there is a regularity that the working hours of unionized workers have tended to be shorter than those of nonunion workers. There are exceptions, but it is a common finding. Cross-sectional differences do not automatically translate into time-series changes, but insofar as gains initially achieved in unionized workplaces establish a new norm that induces nonunion workplaces to follow, then unions have provided the initial stimulus for a pattern over time.

Second, reductions in hours of work over time tend to be concentrated in relatively few years and often in expansionary periods when wages were rising and firms were anxious to take advantage of buoyant product markets. Disruptions in production at that time were particularly costly for firms, but for that reason it was a propitious moment for trade unions to overcome the typical employer's strong opposition to shorter hours. Thus, trade unions exploited business expansions to press their demands for shorter hours, and the two principal goals of unionism—higher wages and shorter hours—came at times when their bargaining effectiveness was at its height.

Third, hours of work have tended to be lower in Britain than in America, and unionism has been a more potent agent in Britain than in America. Not only have a greater fraction of workers been unionized in Britain than in America but also the public standing of unions has been higher in Britain. As agents for workers in the workplace and as a voice in Parliament through their association with the Labour Party, unions have campaigned for better working conditions. They have persuaded large parts of British society that time at work should not jeopardize the well-being of workers and that rising incomes should be spent, in part, on fewer hours and days at the workplace.

The cost in terms of lost output from shorter hours and days has been exaggerated by many employers, legislators, and economists. The unions' campaigns induced some employers to experiment with shorter hours, and these firms did not collapse but, rather, thrived; this, in turn, inspired legislators and economists to alter their conception of what constituted the standard working day and working week.

In a sense, the common explanations for the decline in working hours—the activities of trade unions, the effects of statutory legislation, and the workings of labor markets—are not alternative and competitive but, rather, complementary explanations. Through their trade unions, through their political representatives, and through their actions in choosing to work in firms with more attractive conditions, workers have expressed their preferences for shorter hours of work. The growing realization that long

DIMINISHING RETURNS AT WORK

working hours are not productive ultimately has permitted these preferences to be satisfied.

Because of the ultimate time constraint on all of us, longer hours at the workplace result in less time for other things—less time for sleep and rest, for nourishment, and for the duties of citizenship. Biddle and Hamermesh (1990) quantify the negative association between time at work and time at sleep. Is there a similar negative association between time at work and time devoted to civic duties, such as performing volunteer work or time doing jury service or voting? If so, long working hours entail social and private costs. When there is less time for sleep and certain other activities (such as health maintenance), the performance of workers at their places of work may well be impaired. The pages in this book have provided evidence of this. The marginal product of hours worked is not a constant but, rather, varies with hours: when the typical worker has worked 30 hours, one more hour of work generates more output than if he has worked 40 hours. This is simply the law of diminishing returns as applied to hours of work.

It gives rise to a distinction between nominal hours of work (the hours that individuals are observed to be working in the market) and effective hours of work (the hours that are effective in producing goods and services). The relation between nominal and effective hours is mediated by effort per hour, which is not a constant and which tends to decline as nominal hours lengthen. Another way of expressing this is that nominal hours are discounted by a factor that tends to increase as working hours lengthen.

Two different expressions have been posited for the manner in which effort changes as nominal hours increase. These imply two different specifications of the production function. The estimates of these production functions both confirm the law of diminishing returns as applied to working hours. Neither the hypothesis of unit returns to hours nor the hypothesis of constant returns to hours is sustained. The elasticity of output with respect to hours of work declines as the number of hours increases or, equivalently, the ratio of the marginal product of hours to the average product of hours tends to fall as hours lengthen.

These two production functions have been named "augmented" because working hours enter them in an extended form. These augmented forms represent a notably superior description of the association between hours and output compared with their nonaugmented forms. Both the augmented Cobb-Douglas form and the augmented Gompertz form imply that increases in hours have weak effects on output when hours are "long." Although some might claim that these weak effects arise because the method for estimating them—least-squares—tends to be sensitive to outlier values of hours, similar inferences were drawn when the equations were fitted to subsets of hours where outliers were less marked and influential (as in the results reported in tables 4.5 and 4.13) and to bodies of data where such outliers were not present.

The distinction between nominal and effective hours of work also calls into question the common practice of measuring the input of labor by adding up the nominal hours of all workers or, equivalently, by multiplying average (nominal) hours by the number of workers. Because effort per hour declines with hours, the disparity between nominal hours and effective hours is greater at 60 hours than at 40 hours. Hence, a work force of 120 individuals with each working 40 hours a week may well represent more effective labor than 80 individuals with each working 60 hours a week. The hypothesis that these two combinations of employment and hours per worker that yield the same worker-hours represent equivalent labor services should not be presumed. The effect on output of variations in hours should be distinguished from the effect of variations in employment.

The declining effectiveness of long hours is manifested not only in marketable output but also in a rising probability of ill-health and accidents. Evidence of this has been presented for both blue-collar workers and white-collar workers.

There is not a precise hour that separates "long" hours from "short" hours. Continuous relationships have been estimated that do not specify an abrupt distinction between "long" and "short" hours. Moreover, the relationships vary from plant to plant and from one group of workers to another. This frustrates the imposition of any rule or regulation regarding hours of work that might be applied to all workplaces; production functions differ across workplaces and across workers.

Many economists have already accepted (at least, tacitly) the law of diminishing returns as applied to hours of work, so they may wonder why many pages have been devoted to empirical research that demonstrates the law of diminishing returns applies to working hours. The answer is that, when the law has been accepted, often it has been accepted unthinkingly with little consideration of its implications. Diminishing returns to working hours implies that, at the hours common in the nineteenth century, a mandatory 10% reduction in hours would have resulted in less than a 10% reduction in output, contrary to the claims made by many employers and by economists at the time, so that the resources spent by employers to resist hours reductions were not spent wisely. Diminishing returns to working hours implies that, at times, British munitions workers in the First World War were endangering their health, and in some instances their lives, for meager material gain. Diminishing returns to working hours implies that, today, for workers in some occupations, working hours are at levels that have damaging health consequences for them. Diminishing returns to working hours implies that, in claiming to estimate the wage–working hours preferences of workers, economists have neglected an identification problem in interpreting the association between working hours and hourly earnings.

If the workers' supply function of hours were correctly identified, my suspicion is that it would differ by type of work. Where the work is physically or mentally fatiguing and stressful, and where the hours are long, the supply of hours is likely to be negatively sloped with respect to hourly wages. I would not expect the slope to be such that weekly earnings would fall as hourly pay rises, so that the elasticity of weekly hours with respect to hourly earnings lies between zero and minus unity. The plywood workers analyzed in chapter 4 provide an example.

Where the work is less disagreeable and provides some satisfaction, the supply of hours may well be positively sloped with respect to hourly wages. Indeed, an upward-sloping supply curve is consistent with the growth in recent decades in long hours worked by college-educated workers (as documented in chapter 1): the relative demand for skilled workers has increased, with the consequence of raising the pay, the employment, and the hours of these workers.

Whether the supply of hours is positively or negatively sloped with respect to hourly earnings, I suspect the labor-supply function exhibits smaller (in absolute value) hours changes in response to wage changes than those implied by employers' negatively sloped demand for hours functions. Substitution possibilities among inputs allows employers to make larger hours their responses to wage changes.

The judgment here that the resistance of employers to reductions in hours was based on an incomplete understanding of their own production functions has also been made by others.

Writing of the period from 1856 to 1873 in Britain, Matthews et al. (1982) assume the reduction in average weekly hours worked from 65 to 56 left output completely unaffected, from which they infer that "employers had previously been irrational or ill-informed in insisting on longer hours. We do not regard this as an objection to the assumption" (104). Similarly, Hicks (1963) noted that, after hours of work were reduced, "employers had been working at more than the optimum without realising it" (107).

This conclusion about employers will provoke, in some quarters, a response similar to that offered by Henry Fawcett (Professor of Political Economy at Cambridge University) in 1863, in opposition to proposed statutory restrictions on the hours of work of adults: "Are not the circumstances peculiar to each trade best known to those who are engaged in it, and are they not, consequently, in a far better position to judge the number of hours of labour appropriate to it than the heterogeneous assembly called the State?" (1872, 119–120).

As Professor Fawcett must have known, those who were engaged in each trade tended to have divergent views on the appropriate number of hours of work. As expressed by trade unions, workers sought to reduce hours without a cut in weekly earnings. In resisting these pressures, employers claimed that hours reductions would have disastrous[1] consequences for them and for the economy. Unfortunately, neither employers nor economists such as Professor Fawcett spelled out the mechanism by which a cut in hours would have such damaging effects, and insofar as

---

1. The word "disastrous" was used by David M. Parry in testifying before the US Senate Committee on Education and Labor in 1904, regarding proposed legislation to require government contractors to limit the working day to 8 hours. Parry was the president of the National Association of Manufacturers. Another businessman used the word "calamitous."

Concluding Notes

their opposition rested on their understanding of the manner in which hours enter the production function, it was based on the denial of a proposition that today is widely accepted: the law of diminishing returns.

A rejoinder to this argument might take the form of the following: if shorter hours would not have had such damaging effects for owner-managers in the nineteenth century, then the owner-manager who stumbled upon this would demonstrate the resultant benefits of shorter hours, enjoy commercial success, and other employers would be induced to mimic the practice. In fact, some employers *did* institute shorter hours—Robert Owen, William Mather, Henry Ford, W. K. Kellogg, George and Richard Cadbury, Joseph and Seebohm Rowntree, Cyrus and James Clark—and they *did* enjoy commercial success. However, the response of other owner-managers was to dismiss these experiences as not reproducible, and they did not mimic their practice.

By no means was Professor Fawcett the only economist who advised against constraints on the hours of adult workers. As historians of economic doctrine have noted, many other economists in the nineteenth century expressed opinions similar to Fawcett's. These opinions were not based on empirical research that examined the effects of reductions in hours but, instead, on the Panglossian conviction that no improvement on the order of things was possible and it was best to leave things as they were. This ideology served the interests of those many employers resisting pressures to reduce working hours. The prevailing dogma of economics provided the intellectual justification for employers' opposition to mandatory shorter working hours.

Although some may express disbelief at the idea, the historical experience is consistent with the notion that it is the individual worker, the one who actually undertakes the tasks at hand, who is likely to know how much time is required to complete the tasks satisfactorily, and that the employer or plant manager is in an inferior position to determine this. While the employer or sales manager may be better informed than workers of product market conditions, with respect to production, there is an inequality of information in which the employer may be in an inferior position. This inequality may help to explain why the worker co-ops in Washington State's plywood industry were estimated in chapter 4 to produce about 20% more from measured inputs than conventional capital-owned mills: the workers knew the production process better than the supervisors and managers in the conventional mills.

Today there are sectors of the American and British economies in which long working hours are routine. Indeed, these hours are praised as indicating a worker's extraordinary contribution to an organization. However, how often is that contribution actually measured rather than presumed? It is the presumption that has been questioned in this book: long hours do not necessarily translate into an extraordinary contribution.

Once the notion is permitted that owner-managers may not be fully informed of their own production functions, constraints on working hours set down in statute law or in union-negotiated collective bargaining agreements such as those specifying the maximum length of the working day and working week, and penalty rates of pay for each hour worked beyond a certain length, may not be viewed as interference in management prerogatives but, rather, as judicious forms of elevating the well-being and perhaps the efficiency of the workplace.

If employers or their plant managers do not know the effects of different working hours, why can they not follow William Mather's example in 1892 and conduct experiments? By reducing their labor costs, the first-order effect of an experimental cut in hours is to raise profits. It is possible that such experiments might discover that longer hours are warranted in some instances. Perhaps government might use tax incentives to encourage such experiments. What is appropriate in one organization may be quite different for another.

Might an owner-manager find benefits in following Seebohm Rowntree's example and consult (perhaps by ballot) employees on their preferences for hours of work? As the marginal product of hours of part-time workers has been shown to exceed that of full-time workers, should employers lean toward substituting part-time jobs for full-time jobs?

The distinction between nominal hours and effective hours that arises from consideration of work effort may be useful in other contexts as well. In the research reported here, effort per hour depends upon (nominal) hours of work. It is likely, however, that effort varies with other observables—the characteristics of the workers (such as their age), characteristics of their workplaces (such as the extent to which the capital is complementary with labor), and characteristics of the larger environment (such as the temperature and humidity). Indeed, the empirical work in this book has covered only a few workplaces in merely two economies and has entertained only the hypothesis that work effort changes with working hours. It would be instructive if empirical work investigated whether the inferences here extend to other industries and other economies. There is still much to learn about working hours.

Data Appendix

VERNON'S WEEKLY OBSERVATIONS

The data that follow consist of 124 weekly observations on hours scheduled and hours actually worked by a given group of workers and their average output in that week. The label "Week ending x/y/z" means the xth day of the y month in the year 19z. When seven days are worked in the week, x is a Sunday. When six days are worked in the week, x is a Saturday.

The source of most of these observations is the Health of Munition Workers Committee (HMWC), of the Ministry of Munitions, Memorandum No. 18, Appendix to Memorandum No. 5 (Hours of Work), Further Statistical Information Concerning Output in Relation to Hours of Work with Special Reference to the Influence of Sunday Labour. For the 27 weekly observations on 40 women milling a screw thread, the 32 weekly observations on 56 men sizing fuze bodies, and the 9 weekly observations on 15 youths boring top caps, Memorandum No. 18 provides the sole source.

The largest number of observations consist of the 56 weekly observations on 100 women turning fuze bodies. For these women, the observations from the week ending May 13, 1916, to the week ending December 23, 1916 (32 weeks) is also from Memorandum No. 18, but the

Appendices

data from the week ending November 14, 1915, to April 23, 1916, are from Memorandum No. 12 or, fully, Health of Munition Workers Committee, Ministry of Munitions, Memorandum No. 12, Appendix of Memorandum No. 5 (Hours of Work), Table I, Statistical Information Concerning Output in Relation to Hours of Work. Unfortunately, Vernon slightly altered one key definition for the observations in Memorandum No. 18 compared with those in Memorandum No.12. Therefore, I revised his data for these women from Memorandum No. 12 to make them comparable with those in Memorandum No.18. The procedure I followed for this revision is laid out below.

In Memorandum No. 18, Vernon writes in paragraph 2, "The actual hours recorded in this and subsequent Tables differ from those quoted in the previous Memorandum [No. 12] for they take account of absent workers and those putting in short weeks." In other words, if:

H^N = scheduled (what Vernon calls "nominal" hours) weekly hours
H^B = hours of "broken" time
H^A = hours lost to absenteeism
H^S = hours lost to workers putting in short weeks
H = actual hours worked,

then, in Memorandum No. 12, $H = H^N - H^B$: call this H_1 and this H_1 was multiplied by the index of output per hour (call this I) to get weekly output.

By contrast, in Memorandum No. 18, $H = H^N - H^B - H^A - H^S$: call this H_2, which was also multiplied by the index of output per hour to get weekly output.

"Hours scheduled differ from hours actually worked because of sickness, unexplained absence, and time spent in setting up before production and cleaning up after spells of production." To place H_1 on the same basis as H_2, Vernon reports that in the weeks in 1915, $H^A + H^S$ was 3.4% of nominal hours while, in 1916, $H^A + H^S$ was 11.1% of nominal hours. Therefore, in the weeks of 1915, H^N was reduced by 0.034 and in the weeks in 1916 H^N was reduced by 0.111 before deducting broken hours. The result was then multiplied by the output index to arrive at weekly output.

Data Appendix

This adjustment was not used for the analysis reported in Pencavel (2015a) and (2015b).

In the observations that follow:

$H(j,t)$ = average hours worked in week ending t by workers belonging to group j
$H^N(j, t)$ = hours scheduled in week ending t of the j workers
$X(j, t)$ = average output of group j workers in week ending t
$S(j, t)$ = 1 if group j workers worked on Sunday in week ending t, = 0 otherwise

There are four groups of workers:

56 weekly observations on 100 women turning fuze bodies
27 weekly observations on 40 women milling a screw thread
32 weekly observations on 56 men sizing fuze bodies
9 weekly observations on 15 youths boring top caps

In total, there are 124 weekly observations: 56 + 27 + 32 + 9.

Observations numbered 1 to 56 are on 100 women turning fuze bodies, observations numbered 57 to 83 are on 40 women milling a screw thread; observations numbered 84 to 115 are on 56 men sizing fuze bodies; and observations numbered 116 to 124 are on 15 youths boring top caps.

100 WOMEN TURNING FUZE BODIES (MODERATELY HEAVY LABOR)

56 Weekly Observations from November 14, 1915, to December 23, 1916

The observations on these women are taken from: Health of Munition Workers Committee, Ministry of Munitions, Memorandum No. 12, Appendix of Memorandum No. 5 (Hours of Work), Table I, Statistical Information Concerning Output in Relation to Hours of Work and from Ministry of Munitions, Health of Munition Workers Committee, Memorandum No. 18, Appendix to Memorandum No. 5 (Hours of Work),

Appendices

Further Statistical Information Concerning Output in Relation to Hours of Work with Special Reference to the Influence of Sunday Labour, Table I.

40 WOMEN MILLING A SCREW THREAD ON THE FUZE BODIES (WOMEN ENGAGED IN LIGHT LABOR)

27 Weekly Observations from December 19, 1915, to November 18, 1916
From Health of Munition Workers Committee, Ministry of Munitions, Memorandum No. 18, Appendix of Memorandum No. 5 (Hours of Work), Further Statistical Information Concerning Output in Relation to Hours of Work with Special Reference to the Influence of Sunday Labour, Table II.

FIFTY-SIX MEN SIZING FUZE BODIES (HEAVY LABOR)

32 Weekly Observations from December 19, 1915, to December 23, 1916
From Health of Munition Workers Committee, Ministry of Munitions, Memorandum No. 18, Appendix of Memorandum No. 5 (Hours of Work), Further Statistical Information Concerning Output in Relation to Hours of Work with Special Reference to the Influence of Sunday Labour, Table III.

FIFTEEN YOUTHS BORING TOP CAPS

9 Weekly Observations from December 19, 1915, to December 16, 1916; Youths ages 15 to 18 years
From Health of Munition Workers Committee, Ministry of Munitions, Memorandum No. 18, Appendix of Memorandum No. 5 (Hours of Work), Further Statistical Information Concerning Output in Relation to Hours of Work with Special Reference to the Influence of Sunday Labour, Table IV.

SUNDAY LABOUR

In Health of Munition Workers Memorandum No. 18, Appendix to Memorandum No. 5 (Hours of Work), H. M.Vernon writes on page 4 paragraph 5: "The weeks in which Sunday labour was performed are indicated in the Table[s] by putting normal [scheduled] hours of work in italics." This is done for each of the tables in this memorandum.

Data Appendix

100 Women Turning Fuze Bodies from 14 November, 1915, to 23 December, 1916

56 Weekly Observations

obs	Week ending	Weekly hours		Index of total output $X(t)$	$S(t)$
		actual $H(t)$	scheduled $H^N(t)$		
1	14/11/15	59.7	67.5	5850.6	1
2	21/11/15	66.2	75.5	6553.8	1
3	28/11/15	64.2	75.0	6548.4	1
4	5/12/15	68.4	77.2	6566.4	1
5	12/12/15	66.5	76.2	6583.5	1
6	19/12/15	69.2	77.3	7404.4	1
7	26/12/15	40.2	46.0	4221	1
8	2/1/16	31.6	35.5*	2812.4	1
9	9/1/16	62.8	69.3	7096.4	1
10	16/1/16	67.7	77.2	7243.9	1
11	23/1/16	67.7	76.3	7582.4	1
12	30/1/16	60.1	68.5	6671.1	1
13	6/2/16	58.5	66.5	5967	1
14	13/2/16 ▲	47.4	52.0	5151.6	1
15	20/2/16 ▲	45.8	52.0	4854.8	1
16	27/2/16	54.0	66.5	6372	1
17	5/3/16	54.8	66.5	6850	1
18	12/3/16	48.5	58.7	6159.5	1
19	19/3/16	54.7	66.5	6618.7	1
20	26/3/16	53.0	66.5	6413	1

Note: ▲ denotes weeks when hours were reduced because of temporary shortage of material.

* *inferred*: The value for scheduled hours for the week ending 2/1/16 is missing. I formed the ratio of scheduled hours to actual hours in the previous week (i.e., 46.0 ÷ 41.8 = 1.100) and in the subsequent week (i.e., 69.3 ÷ 65.2 = 1.063) and constructed the average of 1.100 and 1.063 (1.0815) and estimated scheduled hours for 2/1/16 as the product of actual hours in 2/1/16 (32.8) and 1.0815 which is 35.5.

100 Women Turning Fuze Bodies (continued)

obs	Week ending	Hours		Output $X(t)$	$S(t)$
		actual $H(t)$	scheduled $H^N(t)$		
21	2/4/16	51.4	64.8	6219.4	1
22	9/4/16	48.4	58.5	5856.4	1
23	16/4/16	55.5	66.5	6993	1
24	23/4/16	41.5	49.5	5187.5	1
		Easter break			
25	13/5/16	55.8	58.5	7,254.0	0
26	21/5/16	58.5	66.5	7,488	1
27	28/5/16	56.1	66.5	7,405.2	1
28	4/6/16	59.1	66.5	8,037.6	1
29	10/6/16	51.0	58.5	6,987	0
30	17/6/16	50.3	58.5	7,142.6	0
31	24/6/16	53.1	58.5	7,062.3	0
32	1/7/16	54.2	58.5	7,208.6	0
33	8/7/16	56.0	62.5	7,056.0	0
34	15/7/16	48.5	52.5	6,499.0	0
35	22/7/16	45.6	51.0	6,156.0	0
36	30/7/16	56.2	66.5	7,193.6	1
37	6/8/16	48.6	62.5	5,929.2	1
38	12/8/16	40.3	49.5	5,077.8	0
39	19/8/16	43.0	49.5	5,203.0	0
	26/8/16	holiday			

Data Appendix

100 Women Turning Fuze Bodies (continued)

obs	Week ending	Hours		$X(t)$	$S(t)$
		actual $H(t)$	scheduled $H^N(t)$		
40	2/9/16	48.6	58.5	6,609.6	0
41	9/9/16	51.3	58.5	6,822.9	0
42	16/9/16	52.6	58.5	6,771.2	0
43	23/9/16	47.2	56.5	6749.6	0
44	30/9/16	24.0	29.5	3,264.0	0
45	7/10/16	43.5	58.5	5,959.5	0
46	14/10/16	49.2	58.5	7,183.2	0
47	21/10/16	52.0	58.5	7,956.0	0
48	28/10/16	46.9	54.5	6,190.8	0
49	4/11/16	50.0	58.5	7,600.0	0
50	11/11/16	44.1	53.5	6,306.3	0
51	18/11/16	48.8	58.5	6,880.8	0
52	25/11/16	41.8	50.0	6,520.8	0
53	2/12/16	41.6	49.5	6,323.2	0
54	9/12/16	48.8	58.5	8,491.2	0
55	16/12/16	48.8	58.5	8,735.2	0
56	23/12/16	24.5	29.5	3,650.5	0

Appendices

40 Women Milling a Screw Thread, 27 weeks

obs	Week ending	Hours		Output $X(t)$	$S(t)$
		actual $H(t)$	scheduled $H^N(t)$		
57	19/12/15	64.9	71.8	6490.0	1
58	16/4/16	55.4	62.9	6038.6	1
59	28/5/16	62.5	66.5	7062.5	1
60	4/6/16	63.1	66.5	7004.1	1
61	10/6/16	46.6	55.0	5498.8	0
62	17/6/16	46.3	55.0	5231.9	0
63	24/6/16	51.9	56.5	6124.2	0
64	2/7/16	56.8	66.5	6475.2	1
65	9/7/16	54.8	62.0	5973.2	1
66	16/7/16	59.8	66.5	7056.4	1
67	23/7/16	55.6	66.5	6783.2	1
68	30/7/16	53.5	66.5	6366.5	1
69	6/8/16	50.5	59.0	6413.5	1
70	12/8/16	29.5	33.5	3628.5	0
71	19/8/16	38.1	44.5	4648.2	0
	26/8/16		week's holiday		
72	2/9/16	51.1	57.0	6183.1	0
73	9/9/16	47.5	51.1	5985.0	0
74	16/9/16	38.5	44.5	4504.5	0
75	23/9/16	45.0	48.5	5445.0	0
76	30/9/16	26.4	29.5	3141.6	0
77	7/10/16	44.8	54.5	4928.0	0
78	14/10/16	49.3	54.5	6507.6	0
79	21/10/16	48.7	58.5	6428.4	0
80	28/10/16	49.5	55.5	6880.5	0
81	4/11/16	49.9	58.5	6487.0	0
82	11/11/16	43.1	58.5	5775.4	0
83	18/11/16	48.3	58.5	6230.7	0

Data Appendix

32 Weekly Observations on 56 Men Sizing Fuze Bodies

obs	Week ending	Hours		Output X(t)	S(t)
		actual H(t)	scheduled $H^N(t)$		
84	19/12/15	58.2	66.7	5820.0	1
85	16/4/16	50.5	62.8	6161.0	1
86	28/5/16	54.9	66.5	6423.3	1
87	4/6/16	55.8	63.5	6807.6	1
88	10/6/16	49.9	58.5	5888.2	0
89	17/6/16	46.0	53.5	5474.0	0
90	24/6/16	51.2	58.5	6246.4	0
91	1/7/16	48.3	58.5	5699.4	0
92	9/7/16	51.7	61.0	5790.4	1
93	16/7/16	59.1	66.5	7092.0	1
94	22/7/16	47.6	55.5	5664.4	0
95	29/7/16	45.9	58.5	5691.6	0
96	5/8/16	45.3	53.2	5707.8	0
97	12/8/16	27.8	31.0	3614.0	0
98	19/8/16	33.0	39.5	4488.0	0
	26/8/16		holiday		

56 Men Sizing Fuze Bodies, continued

obs	Week ending	Hours		Output X(t)	S(t)
		actual H(t)	scheduled $H^N(t)$		
99	2/9/16	51.6	58.5	6759.6	0
100	9/9/16	49.2	58.5	6642.0	0
101	16/9/16	40.1	44.5	5333.3	0
102	23/9/16	49.4	53.5	6866.6	0
103	30/9/16	28.2	29.5	3807.0	0
104	7/10/16	51.3	54.5	6822.9	0
105	14/10/16	51.4	54.5	7298.8	0
106	21/10/16	50.2	55.3	6927.6	0
107	28/10/16	47.7	51.8	6344.1	0

Appendices

obs	Week ending	Hours		Output X(t)	S(t)
		actual $H(t)$	scheduled $H^N(t)$		
108	4/11/16	55.7	58.5	7686.6	0
109	11/11/16	53.3	58.5	7355.4	0
110	18/11/16	50.6	56.0	6881.6	0
111	25/11/16	50.7	55.2	6844.5	0
112	2/12/16	49.1	54.5	6874.0	0
113	9/12/16	50.6	58.5	7134.6	0
114	16/12/16	53.0	58.5	7738.0	0
115	23/12/16	27.1	29.5	3685.6	0

9 Weekly Observations on 15 Youths Boring Top Caps

obs	Week ending	Hours		Output X(t)	S(t)
		actual $H(t)$	scheduled $H^N(t)$		
116	19/12/15	72.5	78.5	7,250.0	1
117	13/2/16	69.1	75.5	7,324.6	0.67
118	16/4/16	54.8	63.4	5918.4	0.75
119	28/5/16	54.7	61.5	6399.9	0.50
120	1/7/16	47.4	51.1	5877.6	0
121	23/9/16	52.8 day	59.7	6,811.2	0.33
122	23/9/16	56.2 night	61.6	7,193.6	0
123	16/12/16	48.7 day	54.8	6,136.2	0
124	16/12/16	54.7 night	58.2	6,892.2	0

71 Monthly Observations from the Industrial Health Research Board (IHRB)

There are six groups of workers:
  Group 1 (11 monthly observations) consists of 115 women workers
  Group 2 (12 monthly observations) consists of 75 men
  Group 3 (12 monthly observations) consists of 190 men
  Group 4 (12 monthly observations) consists of 190 men
  Group 5 (12 monthly observations) consists of 200 men
  Group 6 (12 monthly observations) consists of 200 men

In the table that follows, 4/40 under "month/year" means the fourth month (April) in 1940.

H_{jt} is the weekly hours of work averaged over the weeks in the month t of Group j workers.

X_{jt} is the weekly output averaged over the weeks in the month t of Group j workers.

Appendices

Month/year	H	X	Group	Obs.
4/40	53.4	5340	1	1
6/40	61.2	6089	1	2
7/40	59.2	5701	1	3
8/40	51.2	4675	1	4
9/40	44.1	4331	1	5
10/40	49.2	4718	1	6
11/40	52.4	4795	1	7
12/40	47.5	4408	1	8
1/41	53.6	5274	1	9
2/41	54.3	5376	1	10
3/41	55.0	5522	1	11
4/40	48.8	4880	2	12
6/40	48.2	4550	2	13
7/40	47.3	4517	2	14
8/40	42.9	3874	2	15
9/40	42.9	3852	2	16
10/40	42.7	4048	2	17
11/40	41.7	4224	2	18
12/40	36.2	3566	2	19
1/41	45.0	4603	2	20
2/41	46.4	4719	2	21
3/41	46.2	4726	2	22
4/41	42.9	4118	2	23
4/40	40	4000	3	24
6/40	46.4	4598	3	25
7/40	41.2	4309	3	26
8/40	36.8	3790	3	27
9/40	36.5	3927	3	28
10/40	42.1	4433	3	29
11/40	38.3	3727	3	30
12/40	34.6	3239	3	31
1/41	40.4	4319	3	32
2/41	43.9	4754	3	33
3/41	43.5	4672	3	34
4/41	38.8	4124	3	35
4/40	46.8	4680	4	36

Data Appendix

Month/year	H	X	Group	Obs.
6/40	53.0	5152	4	37
7/40	51.3	5238	4	38
8/40	46.1	4324	4	39
9/40	39.8	4247	4	40
10/40	45.0	4540	4	41
11/40	43.3	4226	4	42
12/40	42.4	4142	4	43
1/41	49.0	4929	4	44
2/41	50.2	4925	4	45
3/41	49.7	5348	4	46
4/41	43.7	4597	4	47
4/40	52.8	5280	5	48
6/40	63.6	6621	5	49
7/40	60.6	6618	5	50
8/40	57.4	6297	5	51
9/40	60.0	6648	5	52
10/40	58.8	6609	5	53
11/40	57.8	6676	5	54
12/40	55.5	6310	5	55
1/41	55.1	5962	5	56
2/41	51.5	5413	5	57
3/41	59.8	6650	5	58
4/41	55.8	6562	5	59
4/40	57.2	5720	6	60
6/40	61.3	6314	6	61
7/40	58.8	6298	6	62
8/40	58.6	6411	6	63
9/40	58.5	6529	6	64
10/40	57.6	6526	6	65
11/40	54.5	6022	6	66
12/40	48.6	5395	6	67
1/41	54.4	5973	6	68
2/41	53.8	6359	6	69
3/41	53.1	6648	6	70
4/41	50.4	6668	6	71

92 Observations from Kossoris and Kohler

In *Hours of Work and Output*, US Department of Labor, Bureau of Labor Statistics, Bulletin no. 917, 1948.

In the table that follows, ND means not determined and OBS denotes the observation number.

13 observations (OBS) numbered 1 to 13 correspond to a situation when daily and weekly hours were increased but the weekly days worked remained constant at 5 (from K&K, tables 1 and 2, pp. 9–10).

15 observations (OBS) numbered 14 to 28 correspond to a situation when daily hours were constant but working days increased from 5 to 6 (from K&K, tables 3 and 4, pp. 13–14).

14 observations (OBS) numbered 29 to 42 correspond to a situation when daily hours increased and weekly working days increased from 5 to 6 (from K&K, tables 7 and 8, pp. 18–20).

22 observations (OBS) numbered 43 to 64 correspond to a situation when daily and weekly hours were increased but the weekly days worked remained constant at 6 (from K&K, tables 12 and 13, pp. 24–26).

4 observations (OBS) numbered 65 to 68 correspond to a situation when daily and weekly hours were increased but the weekly days worked were reduced from 6 to 5 (from K&K, tables 9 and 10, pp. 21–22).

6 observations (OBS) numbered 69 to 74 correspond to a situation when weekly hours were unchanged but the weekly days worked fell from 6 to five-and-a-half (from K&K, table 11, p. 23).

Data Appendix

4 observations (OBS) numbered 75 to 78 correspond to a situation when daily and weekly hours were reduced but days worked remained unchanged at 6 (from K&K, table 14, p. 27).

14 observations (OBS) numbered 79 to 92 correspond to a situation when daily hours were unchanged but days worked fell from 6 to 5. The percent change in output reported by K&K is computed from the lower level of hours "as though hours had increased from this level rather than reduced to it so as to make the results comparable with other tables" (pp. 16 and 21) (from K&K, tables 5 and 6, pp. 15–66).

Definitions of variables that follow, from K&K 1948:

H_B denotes weekly hours worked **B**efore the change in working schedules.

H_A denotes weekly hours worked **A**fter the change in working schedules.

Hours worked deducts hours lost due to absenteeism from scheduled hours.

D_B denotes the days worked in a week **B**efore the change in work schedules.

D_A denotes the days worked in a week **A**fter the change in work schedules.

ΔX denotes the percentage change in weekly output.

Y_B denotes the absenteeism rate **B**efore the change in working schedules.

Y_A denotes the absenteeism rate **A**fter the change in working schedules.

Absenteeism is "worktime lost because workers did not show up or stay on the job when they were scheduled to work" (p. 60). The absenteeism rate is the total employee-hours lost in this way as a percentage of total employee-hours scheduled.

N denotes the number of workers covered by the change in work schedules.

W indicates the gender of the workers: $W = 1$ indicates women workers; $W = 0$ indicates men workers; and $W = BOTH$ indicates both women and men workers.

$L = 1$ if the work is designated as "light" as distinct from "heavy" or "moderately heavy."

$C = 1$ if the work is controlled by the operator as distinct from the machine.

Appendices

OBS	Weekly hours H_B	H_A	Days worked D_B	D_A	ΔX	Absenteeism rate Y_B	Y_A	N	W	L	C
1	40	50	5	5	23.5	2.7	7.0	39	0	0	0
2	40	50	5	5	18.0	5.4	7.0	50	0	0	0
3	40	45	5	5	6.4	ND	ND	13	1	1	1
4	40	48	5	5	0.6	ND	ND	13	1	1	1
5	40	46.5	5	5	13.3	3.2	4.6	24	1	1	1
6	44.5	46.5	5	5	0.2	2.8	2.8	10	1	1	1
7	40	46.5	5	5	11.8	5.4	3.1	10	1	1	1
8	40	46.5	5	5	14.2	3.1	2.1	21	1	1	1
9	40	44	5	5	6.7	3.1	2.5	21	1	1	1
10	40	44	5	5	7.9	1.9	0.8	13	1	1	1
11	40	47.5	5	5	18.6	5.5	4.3	13	1	1	1
12	40	43	5	5	-1.1	8.4	16.7	28	1	1	1
13	45	50	5	5	11.8	5.3	5.8	9	1	1	1
14	50	58	5	6	10.1	7.0	4.5	39	0	0	0
15	50	58	5	6	10.2	7.0	8.1	50	0	0	0
16	40	48	5	6	21.0	ND	ND	700	0	0	0
17	40	44.5	5	6	5.9	1.2	4.2	32	0	0	1
18	40	48	5	6	22.8	0.7	1.7	21	0	0	1
19	40	48	5	6	21.4	1.1	1.4	8	0	0	1
20	40	48	5	6	20.8	1.7	2.8	10	0	0	1
21	40	48	5	6	22.8	3	4.6	14	0	0	1
22	50	58	5	6	15.1	6.8	7.7	10	1	1	1
23	40	48	5	6	22.8	1.4	0.5	9	1	1	1
24	40	48	5	6	14.9	3	6.8	8	1	1	1
25	40	46	5	6	10.7	8.4	10.3	28	1	1	1
26	40	48	5	6	20.2	1	1.7	13	1	1	1
27	40	48	5	6	17.9	2	4.7	9	1	1	1
28	40	48	5	6	19.3	2	9.2	9	1	1	1
29	40	58	5	6	36	2.7	4.5	39	0	0	0
30	40	58	5	6	30	5.4	8.1	50	0	0	0
31	40	60	5	6	5.5	5.7	10.8	65	BOTH	0	1

Data Appendix

	Weekly hours		Days worked			Absenteeism rate					
OBS	H_B	H_A	D_B	D_A	ΔX	Y_B	Y_A	N	W	L	C
32	40	66	5	6	9.0	3.7	13.3	55	BOTH	0	1
33	40	55	5	6	29.6	2.7	2.6	18	0	0	1
34	50	48	5	6	-0.7	3.3	3.1	14	0	0	1
35	40	55	5	6	43.6	1.8	2.0	14	0	0	1
36	40	57	5	6	67.2	3.0	2.9	14	0	0	1
37	37.5	58	5	6	81.2	2.1	3.3	43	0	1	1
38	37.5	54	5	6	67.6	3.0	3.3	25	1	1	1
39	40	48	5	6	11.7	3.0	11.8	8	1	1	1
40	40	54	5	6	22.6	3.0	14.4	8	1	1	1
41	40	53	5	6	13.9	4.5	9.4	24	1	1	1
42	40	48	5	6	3.3	4.5	8.3	24	1	1	1
43	52	58	6	6	3.7	3.3	2.8	48	0	0	1
44	48	55.5	6	6	13.2	6.6	8.9	65	0	0	1
45	48	54	6	6	13.5	6.9	8.5	68	0	0	1
46	45	58	6	6	25.9	4.2	6.9	275	0	0	1
47	48	57	6	6	36.2	4.6	2.9	14	0	0	1
48	45	55	6	6	20.9	3.3	4.5	66	1	1	1
49	45	57	6	6	15.2	3.0	4.7	61	1	1	1
50	45	55	6	6	8.4	2.2	5.8	26	1	1	1
51	45	55	6	6	12.6	3.3	4.1	22	1	1	1
52	45	57	6	6	20.0	3.0	4.7	59	1	1	1
53	45	55	6	6	15.9	1.7	3.6	100	BOTH	1	1
54	45	57	6	6	14.8	3.6	6.1	50	BOTH	1	1
55	46	58	6	6	36.4	4.9	15.4	10	1	1	1
56	46	56	6	6	48.2	5.7	7.7	10	1	1	1
57	46	61	6	6	39.9	3.5	5.7	15	0	1	1
58	46	61	6	6	45.9	3.9	6.7	15	0	1	1
59	46	58	6	6	31.9	2.0	4.2	92	0	1	1
60	54	61.5	6	6	4.2	5.1	6.3	16	0	1	1
61	48	54	6	6	3.9	0.5	3.8	6	0	1	1
62	48	54	6	6	6.7	6.8	14.4	8	1	1	1

Appendices

OBS	Weekly hours		Days worked		ΔX	Absenteeism rate		N	W	L	C
	H_B	H_A	D_B	D_A		Y_B	Y_A				
63	48	54	6	6	17.8	8.0	8.3	15	1	1	1
64	40	44	6	6	7.4	1.9	7.3	40	1	1	1
65	63	60	6	5	−4.1	7.9	6.1	17	0	0	1
66	46	50	6	5	14.9	4.9	8.3	10	1	1	1
67	40	43.5	6	5	2.6	2.2	9.7	26	1	1	1
68	48	35	6	5	−4.1	5.7	4.1	30	0	1	0
69	48	48	6	5.5	9.6	7.5	7.4	8	BOTH	0	1
70	48	48	6	5.5	−1.0	0.5	4.4	9	0	1	1
71	48	48	6	5.5	−2.8	6.8	11.8	8	1	1	1
72	48	48	6	5.5	12.8	22.1	12.5	5	1	1	1
73	48	48	6	5.5	17.4	18.5	5.5	5	1	1	1
74	48	48	6	5.5	6.3	8.0	4.7	15	1	1	1
75	58	52	6	6	0	2.8	4.2	48	0	0	1
76	58	48	6	6	−7.9	10.7	10.4	140	BOTH	0	1
77	53	48	6	6	−3.0	5.5	7.0	32	0	0	1
78	55	48	6	6	−6.7	3.5	3.1	14	0	0	1
79	50	60	6	5	20	ND	ND	45	0	0	1
80	50	58	6	5	0	4.2	4.7	16	0	0	1
81	50	58	6	5	−3.2	5.5	5.0	15	0	0	1
82	50	58	6	5	23.8	3.2	3.6	33	0	0	1
83	59	58	6	5	7.7	3.4	3.6	18	0	0	1
84	50	55	6	5	6.5	3.3	3.5	14	0	0	1
85	43	48	6	5	14.2	10.7	16.7	28	1	1	1
86	43	46	6	5	12.0	10.3	16.7	28	1	1	1
87	40	48	6	5	12.9	10.7	8.4	28	1	1	1
88	40	48	6	5	11.1	12.1	15.2	17	1	1	1
89	40	48	6	5	12.2	13.7	13.8	25	1	1	1
90	40	48	6	5	13.9	9.2	9.4	26	1	1	1
91	40	48	6	5	22.4	6.4	9.6	25	1	1	1
92	40	48	6	5	13.8	10.4	12.4	9	1	1	1

Observations on Pacific Northwest Plywood Mills

There are 170 mill-year observations shown here on 34 plywood mills from 1968 to 1986. The survey collecting information on output was conducted every two years. This is not a balanced set of observations.

Mill No. 5 (Peninsula Plywood) was converted from a co-op to a unionized mill between 1970 and 1972. It is observed as a co-op in 1968 and 1970 and as a unionized mill between 1972 and 1986.

The Mill No. attaches a number to each mill. There are three types of mills, as follows:

1. There are 8 conventional mills whose workers are not covered by a union contract—these are sometimes called "classical" mills. These classical mills areMill Nos. 1, 2, 3, 6, 8, 19, 28, 39.

2. There are 19 conventional mills whose workers are covered by a union-negotiated collective bargaining agreement. These unionized mills are Mill Nos. 5, 7, 9, 10, 12, 14, 15, 16, 17, 18, 20, 24, 27, 30, 33, 34, 35, 36, 37.

3. There are 7 mills that are owned and managed by the workers who work in them. These are called co-op mills. The co-op mills are Mill Nos. 4, 5, 21, 23, 29, 31, 38.

Mill No. 14 (Anacortes) is observed as a unionized mill in 1968, 1970, 1972, 1974, 1976, 1978 and 1980. After conversion to a co-op, it is observed as a co-op in 1984 and 1986. However, values for its inputs of logs are missing, so this mill does not appears as a co-op in the observations that follow.

For these 170 mill-year observations, there are 3 mills with observations in all 10 even-numbered years. These are Mill No. 8 (a classical mill), Mill No. 27 (a unionized mill), and Mill No. 38 (a co-op).

Two mills provide only one observation each: Mill No. 19 (a classical mill) is observed once (in 1980) and Mill No. 35 (a unionized mill) is observed once (in 1980).

The breakdown of mill-year observations by type of mill and year is as follows:

Year	Observations on classical (nonunion) mills	Observations on unionized mills	Observations on co-op mills	Observations on all three types of mills
1968	2	6	5	13
1970	2	8	3	13
1972	2	11	5	18
1974	2	13	2	17
1976	2	13	3	18
1978	4	14	3	21
1980	6	13	4	23
1982	4	10	4	18
1984	5	8	3	16
1986	4	5	4	13
total	33	101	36	170

In the first set of tables that follow, H is average annual hours worked per worker computed as the number of days of operation over the year multiplied by the hours worked per day; N is the number of workers employed in the mill averaged over the weeks in the year; M is the quantity of raw material logs used by the mill in the year. Log inputs are measured in thousands of feet used during a year; X is annual real output measured as the annual aggregate output in square feet of softwood plywood and veneer. Plywood and veneer are aggregated using region-specific current prices and then deflated by a plywood producer price index.

Data Appendix

Table 1 lists the observations on the input–output variables: table 1.A shows observations on the nonunion conventional mills (referred to as classical); table 1.B shows observations on the unionized conventional firms; table 1.C shows observations on the worker co-ops.

Table 2 lists observations on the real hourly wage (w), the real price of logs (r), and the real price of output (p).

TABLE 1.A Input–Output Observations, Nonunion Conventional Mills: Nos. 1–24

OBS	Mill No.	Year	ID	H	N	M	X
1	1	1980	1/80	1880	68	12000	0.02841
2	1	1982	1/82	1488	78	3000	0.00776
3	1	1984	1/84	1760	73	6166	0.02031
4	2	1968	2/68	1600	26	12435	0.01748
5	2	1970	2/70	1600	24	8300	0.01373
6	2	1972	2/72	2000	44	3733	0.03264
7	2	1974	2/74	1864	44	2690	0.00917
8	2	1976	2/76	1600	44	4000	0.00763
9	2	1978	2/78	1600	48	6027	0.01141
10	3	1978	3/78	3500	95	32815	0.07206
11	3	1980	3/80	1544	77	22712	0.04425
12	3	1984	3/84	400	42	5019	0.00625
13	3	1986	3/86	1760	73	25586	0.06335
14	6	1984	6/84	480	9	100	0.00023
15	6	1986	6/86	1200	8	2000	0.00493
16	8	1968	8/68	1760	181	24000	0.09187
17	8	1970	8/70	1760	179	19000	0.09267
18	8	1972	8/72	1920	160	24000	0.11066
19	8	1974	8/74	1440	105	12000	0.04720
20	8	1976	8/76	1920	88	15200	0.06661
21	8	1978	8/78	1800	108	15000	0.06668
22	8	1980	8/80	880	82	5069	0.02328
23	8	1982	8/82	1600	68	7150	0.03408
24	8	1984	8/84	1920	72	13000	0.05853

213

Input–Output Observations, Nonunion Conventional Mills: Nos. 25–33

Obs	Mill No.	Year	ID	H	N	M	X
25	8	1986	8/86	1720	98	18776	0.09060
26	19	1980	19/80	1200	93	2000	0.00429
27	28	1980	28/80	2000	32	13044	0.02841
28	28	1982	28/82	1600	32	10435	0.02070
29	28	1984	28/84	1600	18	10500	0.00937
30	28	1986	28/86	1600	29	10500	0.01974
31	39	1978	39/78	1056	32	802	0.00204
32	39	1980	39/80	1200	37	1337	0.00341
33	39	1982	39/82	960	37	4500	0.00621

TABLE 1.B Input–Output Observations, Unionized Conventional Mills: Nos. 34–59

Obs	Mill No.	Year	ID	H	N	M	X
34	7	1970	7/70	864	142	7588	0.02344
35	7	1972	7/72	2016	197	42795	0.18523
36	7	1974	7/74	2000	209	60000	0.15985
37	10	1970	10/70	2400	151	45100	0.16739
38	10	1972	10/72	2032	147	56265	0.19329
39	10	1974	10/74	2144	224	49338	0.15554
40	10	1976	10/76	1888	219	45327	0.16228
41	10	1978	10/78	2400	227	43119	0.16779
42	10	1980	10/80	2488	192	52043	0.19246
43	10	1982	10/82	1664	204	35512	0.14820
44	10	1984	10/84	2592	204	46855	0.21708
45	10	1986	10/86	2840	172	49959	0.25095
46	12	1968	12/68	2000	235	52968	0.18894
47	12	1970	12/70	1912	220	46737	0.17704
48	12	1972	12/72	2000	220	55577	0.21405
49	12	1974	12/74	1976	190	51468	0.16341

Data Appendix

Obs	Mill No.	Year	ID	H	N	M	X
50	15	1974	15/74	2000	160	3800	0.02432
51	15	1976	15/76	2000	135	21600	0.09358
52	15	1978	15/78	2000	208	14400	0.06087
53	16	1976	16/76	1976	308	8708	0.13699
54	16	1978	16/78	1920	355	6941	0.14523
55	16	1980	16/78	1600	306	3725	0.10764
56	16	1982	16/82	1584	258	3488	0.09578
57	17	1972	17/72	1920	144	12000	0.03487
58	17	1974	17/74	1760	138	10000	0.02752
59	17	1976	17/76	2000	100	9000	0.03339

Input–Output Observations, Unionized Conventional Mills: Nos. 60–84

Obs	Mill No.	Year	ID	H	N	M	X
60	17	1978	17/78	2080	211	8000	0.01894
61	17	1980	17/80	1496	162	6309	0.01871
62	17	1982	17/82	1336	154	5846	0.00941
63	18	1976	18/76	2016	195	40588	0.14817
64	18	1978	18/78	1984	246	50027	0.18109
65	18	1980	18/80	1832	217	46405	0.16844
66	18	1982	18/82	1208	163	26280	0.10512
67	18	1984	18/84	2280	266	56868	0.23415
68	18	1986	18/86	2016	301	69774	0.26918
69	20	1968	20/68	1224	435	17000	0.10672
70	20	1970	20/70	1888	230	7903	0.04254
71	20	1972	20/72	1896	380	16926	0.12466
72	20	1974	20/74	1816	242	6669	0.08181
73	20	1976	20/76	1912	351	11583	0.14518
74	20	1978	20/78	1864	330	10152	0.10849

Appendices

Obs	Mill No.	Year	ID	H	N	M	X
75	20	1980	20/80	1328	304	2443	0.06451
76	20	1982	20/82	1512	209	2978	0.04734
77	20	1984	20/84	1536	320	3486	0.04405
78	24	1968	24/68	2000	224	18000	0.04154
79	24	1970	24/70	2000	96	10000	0.00915
80	24	1972	24/72	1600	213	12000	0.01844
81	24	1976	24/76	1840	247	10000	0.02671
82	24	1978	24/78	1600	203	6000	0.01515
83	24	1980	24/80	1600	136	6000	0.01515
84	24	1982	24/82	1760	86	6000	0.01380

Input–Output Observations, Unionized Conventional Mills: Nos. 85–111

Obs	Mill No.	Year	ID	H	N	M	X
85	24	1984	24/84	1600	157	5000	0.00635
86	24	1986	24/86	1120	150	6000	0.02212
87	30	1974	30/74	1752	466	24000	0.08746
88	30	1976	30/76	1888	594	24619	0.11539
89	30	1978	30/78	1912	603	27920	0.12518
90	37	1972	37/72	1960	147	34800	0.08468
91	37	1974	37/74	1920	144	35000	0.08074
92	37	1976	37/76	2008	77	18775	0.04579
93	37	1978	37/78	1920	96	26000	0.06722
94	37	1980	37/80	1480	124	17000	0.03409
95	5	1972	5/72	2008	432	34000	0.14770
96	5	1974	5/74	1600	424	24881	0.18322
97	5	1976	5/76	2000	431	27027	0.14633
98	5	1978	5/78	1928	432	21073	0.14124
99	5	1980	5/80	1616	300	7019	0.06228
100	5	1982	5/82	1304	246	4445	0.07036

Data Appendix

Obs	Mill No.	Year	ID	H	N	M	X
101	5	1984	5/84	1640	260	5114	0.08552
102	5	1986	5/86	1744	248	5063	0.09826
103	9	1968	9/68	1920	232	19182	0.09486
104	9	1970	9/70	1920	257	7000	0.12356
105	9	1972	9/72	1960	301	18000	0.13833
106	9	1974	9/74	1400	277	6000	0.07946
107	9	1976	9/76	1960	250	7000	0.06957
108	9	1978	9/78	1936	374	12309	0.11484
109	14	1968	14/68	2400	430	33000	0.21896
110	14	1970	14/70	2040	453	29232	0.19291
111	14	1972	14/72	1785	452	30000	0.22286

Input–Output Observations, Unionized Conventional Mills: Nos. 112–134

Obs	Mill No.	Year	ID	H	N	M	X
112	14	1974	14/74	2450	380	24000	0.13374
113	14	1976	14/76	2032	344	23500	0.14034
114	14	1978	14/78	2016	383	25121	0.18543
115	14	1980	14/80	1384	247	11133	0.08756
116	27	1968	27/68	1920	522	55532	0.12716
117	27	1970	27/70	1832	568	55000	0.12356
118	27	1972	27/72	1904	588	56858	0.23003
119	27	1974	27/74	1896	595	42106	0.15909
120	27	1976	27/76	1944	622	40329	0.12669
121	27	1978	27/78	1960	678	43059	0.14306
122	27	1980	27/80	1848	628	40240	0.12466
123	27	1982	27/82	1576	578	29560	0.10221
124	27	1984	27/84	1840	420	37902	0.11158
125	27	1986	27/86	1888	574	34333	0.16468
126	33	1974	33/74	1824	419	14325	0.11363
127	33	1978	33/78	1928	349	7781	0.14527

Obs	Mill No.	Year	ID	H	N	M	X
128	34	1980	34/80	2080	273	28337	0.09442
129	34	1982	34/82	2000	232	36074	0.14913
130	34	1984	34/84	1880	146	41031	0.09713
131	35	1980	35/80	2168	59	22674	0.05222
132	36	1980	36/80	1856	177	30720	0.17940
133	36	1982	36/82	1832	157	28731	0.16927
134	36	1984	36/84	1800	133	27250	0.10758

TABLE 1.C Input–Output Observations, Worker Co-op Mills: Nos. 135–154

Obs	Mill No.	Year	ID	H	N	M	X
135	21	1970	21/70	2088	466	29000	0.20561
136	21	1972	21/72	2080	465	40000	0.18444
137	21	1974	21/74	1480	406	20000	0.07867
138	23	1980	23/80	2400	339	36000	0.17161
139	23	1982	23/82	1952	323	33387	0.14702
140	23	1984	23/84	1960	317	33142	0.15716
141	23	1986	23/86	2400	329	19000	0.18725
142	29	1968	29/68	1760	136	10000	0.07350
143	29	1972	29/72	2120	148	7090	0.07951
144	29	1976	29/76	2016	161	2998	0.08172
145	29	1978	29/78	2016	159	6690	0.08207
146	38	1968	38/68	2168	232	20000	0.14087
147	38	1970	38/70	1960	236	17500	0.13437
148	38	1972	38/72	2000	238	23000	0.15116
149	38	1974	38/74	2520	227	21420	0.10358
150	38	1976	38/76	2200	254	25700	0.14876
151	38	1978	38/78	1920	254	29101	0.15668
152	38	1980	38/80	1896	245	24190	0.13372
153	38	1982	38/82	1920	238	21843	0.13494
154	38	1984	38/84	2048	246	23000	0.13435

Data Appendix

Input–Output Observations, Worker Co-op Mills: Nos. 155–170

Obs	Mill No.	Year	ID	H	N	M	X
155	38	1986	38/86	2008	243	28610	0.14892
156	4	1968	4/68	2000	249	7000	0.11024
157	4	1972	4/72	2320	230	8145	0.13010
158	4	1980	4/80	2040	285	12000	0.11441
159	4	1982	4/82	2400	251	8000	0.10722
160	4	1986	4/86	2496	252	14545	0.12700
161	5	1968	5/68	2240	405	59672	0.17244
162	5	1970	5/70	2920	369	28182	0.15725
163	31	1968	31/68	2360	232	25000	0.10718
164	31	1972	31/72	2744	227	22733	0.11210
165	31	1976	31/76	2000	207	28291	0.11205
166	31	1978	31/78	2568	237	28528	0.13450
167	31	1980	31/80	2160	305	20000	0.08580
168	31	1982	31/82	2288	259	26000	0.16380
169	31	1984	31/84	2000	266	28000	0.12589
170	31	1986	31/86	2000	309	30000	0.15777

TABLE 2.A Price and Wage Observations, Nonunion Conventional Mills: Nos. 1–24

Obs	Mill No.	Year	ID	Real wage, w	Real log price, r	Real output price, p
1	1	1980	1/80	3.7347	107.415	146.044
2	1	1982	1/82	3.5138	107.988	98.954
3	1	1984	1/84	2.7671	83.176	95.604
4	2	1968	2/68	5.6862	86.379	122.659
5	2	1970	2/70	5.6808	79.196	101.061
6	2	1972	2/72	4.8617	93.888	139.673
7	2	1974	2/74	5.6521	138.884	156.228

Obs	Mill No.	Year	ID	Real wage, w	Real log price, r	Real output price, p
8	2	1976	2/76	6.1793	150.858	162.128
9	2	1978	2/78	7.1105	119.359	185.647
10	3	1978	3/78	2.5965	133.190	185.647
11	3	1980	3/80	3.9810	107.415	146.044
12	3	1984	3/84	6.3386	83.176	95.604
13	3	1986	3/86	2.5522	84.879	92.815
14	6	1984	6/84	0.5192	72.820	58.353
15	6	1986	6/86	0.3823	74.461	50.468
16	8	1968	8/68	3.6132	95.977	124.002
17	8	1970	8/70	3.1053	78.335	97.445
18	8	1972	8/72	3.6193	98.682	123.133
19	8	1974	8/74	4.0018	135.497	127.570
20	8	1976	8/76	4.2489	149.684	143.755
21	8	1978	8/78	4.8355	133.190	166.027
22	8	1980	8/80	5.6741	107.415	126.020
23	8	1982	8/82	3.6764	107.988	97.639
24	8	1984	8/84	3.7327	83.176	98.549

Price and Wage Observations, Nonunion Conventional Mills: Nos. 25–33

Obs	Mill No.	Year	ID	Real wage, w	Real log price, r	Real output price, p
25	8	1986	8/86	3.2651	84.879	95.083
26	19	1980	19/80	5.2545	131.735	126.020
27	28	1980	28/80	4.0005	122.413	146.044
28	28	1982	28/82	3.8460	104.526	98.954
29	28	1984	28/84	4.5055	72.820	95.604
30	28	1986	28/86	3.7730	74.461	92.815
31	39	1978	39/78	6.4913	133.190	185.647
32	39	1980	39/80	6.4112	107.415	146.044
33	39	1982	39/82	6.9051	107.988	98.954

Data Appendix

TABLE 2.B Price and Wage Observations, Unionized Conventional Mills: Nos. 34–59

Obs	Mill No.	Year	ID	Real wage, w	Real log price, r	Real output price, p
34	7	1970	7/70	2.4289	85.222	97.445
35	7	1972	7/72	3.2671	93.888	123.133
36	7	1974	7/74	3.1115	143.627	127.570
37	10	1970	10/70	2.5358	69.727	97.445
38	10	1972	10/72	3.6722	79.105	123.133
39	10	1974	10/74	3.4269	108.398	127.570
40	10	1976	10/76	4.6215	126.791	143.755
41	10	1978	10/78	3.8691	119.871	166.027
42	10	1980	10/80	3.3542	85.932	126.020
43	10	1982	10/82	3.113	96.912	97.639
44	10	1984	10/84	3.1626	80.911	98.549
45	10	1986	10/86	2.8653	76.606	95.083
46	12	1968	12/68	3.9841	95.977	124.002
47	12	1970	12/70	3.9411	78.335	97.445
48	12	1972	12/72	3.3026	98.682	123.133
49	12	1974	12/74	4.6209	135.497	127.570
50	15	1974	15/74	4.1268	135.497	127.570
51	15	1976	15/76	4.3282	149.684	143.755
52	15	1978	15/78	3.7594	133.190	166.027
53	16	1976	16/76	4.6318	121.508	143.755
54	16	1978	16/78	3.9082	119.359	166.027
55	16	1980	16/78	4.1455	122.413	126.020
56	16	1982	16/82	4.0448	104.526	97.639
57	17	1972	17/72	7.8709	98.682	139.673
58	17	1974	17/74	5.8382	135.497	156.228
59	17	1976	17/76	6.4203	149.684	162.128

Price and Wage Observations, Unionized Conventional Mills: Nos. 60–84

Obs	Mill No.	Year	ID	Real wage, w	Real log price, r	Real output price, p
60	17	1978	17/78	3.6957	133.190	185.647
61	17	1980	17/80	6.0415	107.415	146.044
62	17	1982	17/82	6.1065	107.988	98.954
63	18	1976	18/76	3.6758	121.508	143.755
64	18	1978	18/78	4.1077	119.359	166.027
65	18	1980	18/80	4.2803	122.413	126.020
66	18	1982	18/82	1.0300	104.526	97.639
67	18	1984	18/84	3.5149	72.820	98.549
68	18	1986	18/86	3.1147	74.461	95.083
69	20	1968	20/68	7.0308	95.977	124.002
70	20	1970	20/70	5.5400	86.943	97.445
71	20	1972	20/72	4.7983	91.890	123.133
72	20	1974	20/74	7.447	138.884	127.570
73	20	1976	20/76	4.4985	150.858	143.755
74	20	1978	20/78	4.5855	119.359	166.027
75	20	1980	20/80	4.5396	131.735	126.020
76	20	1982	20/82	3.7654	112.487	97.639
77	20	1984	20/84	2.7279	72.820	98.549
78	24	1968	24/68	5.4331	95.977	122.659
79	24	1970	24/70	10.0307	86.943	101.063
80	24	1972	24/72	6.2156	91.890	139.673
81	24	1976	24/76	4.2890	150.858	162.128
82	24	1978	24/78	4.6633	119.359	185.647
83	24	1980	24/80	5.8176	131.735	146.044
84	24	1982	24/82	5.8448	112.487	98.954

Data Appendix

Price and Wage Observations, Unionized Conventional Mills: Nos. 85–110

Obs	Mill No.	Year	ID	Real wage, w	Real log price, r	Real output price, p
85	24	1984	24/84	3.6584	72.820	95.604
86	24	1986	24/86	5.3578	72.009	95.083
87	30	1974	30/74	3.9618	135.497	127.570
88	30	1976	30/76	3.3399	149.684	143.755
89	30	1978	30/78	3.1929	133.190	166.027
90	37	1972	37/72	6.1176	98.682	139.673
91	37	1974	37/74	5.3706	135.497	156.228
92	37	1976	37/76	7.2520	149.684	162.128
93	37	1978	37/78	6.5475	133.190	185.647
94	37	1980	37/80	6.4141	107.415	146.044
95	5	1972	5/72	3.5001	93.888	123.133
96	5	1974	5/74	3.9453	138.884	127.570
97	5	1976	5/76	3.8418	150.858	143.755
98	5	1978	5/78	3.5265	119.359	166.027
99	5	1980	5/80	4.3851	131.735	126.020
100	5	1982	5/82	3.7506	112.487	97.639
101	5	1984	5/84	4.7539	72.820	98.549
102	5	1986	5/86	4.1161	72.009	95.083
103	9	1968	9/68	5.8532	95.977	124.002
104	9	1970	9/70	4.9997	78.335	97.445
105	9	1972	9/72	4.7687	98.682	123.133
106	9	1974	9/74	5.5231	135.497	127.570
107	9	1976	9/76	4.4159	149.684	143.755
108	9	1978	9/78	3.8167	133.190	166.027
109	14	1968	14/68	3.8389	95.977	124.002
110	14	1970	14/70	3.3171	85.222	97.445

Price and Wage Observations, Unionized Conventional Mills: Nos. 111–134

Obs	Mill No.	Year	ID	Real wage, w	Real log price, r	Real output price, p
111	14	1972	14/72	4.0875	93.888	123.133
112	14	1974	14/74	2.8847	143.627	127.570
113	14	1976	14/76	3.7569	121.508	143.755
114	14	1978	14/78	4.3294	119.359	166.027
115	14	1980	14/80	4.6476	122.413	126.020
116	27	1968	27/68	3.7424	91.178	122.659
117	27	1970	27/70	3.9881	87.374	101.061
118	27	1972	27/72	4.1366	91.890	123.133
119	27	1974	27/74	3.2407	143.627	127.570
120	27	1976	27/76	4.5766	121.508	143.755
121	27	1978	27/78	4.5357	119.359	166.027
122	27	1980	27/80	4.6357	122.413	126.020
123	27	1982	27/82	4.1652	104.526	97.639
124	27	1984	27/84	4.8863	72.820	98.549
125	27	1986	27/86	3.9661	74.461	95.083
126	33	1974	33/74	3.9303	143.627	127.570
127	33	1978	33/78	4.0151	119.359	166.027
128	34	1980	34/80	4.3863	107.415	126.020
129	34	1982	34/82	4.4660	107.988	98.954
130	34	1984	34/84	4.8561	83.176	95.604
131	35	1980	35/80	5.2992	107.415	146.044
132	36	1980	36/80	4.3366	122.413	126.020
133	36	1982	36/82	4.0365	104.526	97.639
134	36	1984	36/84	5.0963	72.820	98.549

TABLE 2.C Price and Wage Observations, Worker Co-op Mills: Nos. 135–154

Obs	Mill No	Year	ID	Real wage, w	Real log price, r	Real output price, p
135	21	1970	21/70	2.8490	85.222	97.445
136	21	1972	21/72	3.5301	93.888	123.133
137	21	1974	21/74	2.8197	143.627	127.570
138	23	1980	23/80	3.7839	107.415	126.020
139	23	1982	23/82	2.9351	107.988	97.639
140	23	1984	23/84	3.2933	83.176	98.549
141	23	1986	23/86	2.1730	84.879	95.083
142	29	1968	29/68	3.6993	95.977	124.002
143	29	1972	29/72	3.8657	91.890	123.133
144	29	1976	29/76	4.2988	150.858	143.755
145	29	1978	29/78	4.7595	119.359	166.027
146	38	1968	38/68	3.8032	95.977	124.002
147	38	1970	38/70	3.4904	85.222	97.445
148	38	1972	38/72	4.3135	93.888	123.133
149	38	1974	38/74	2.6520	143.627	127.570
150	38	1976	38/76	3.1752	121.508	143.755
151	38	1978	38/78	4.4560	119.359	166.027
152	38	1980	38/80	2.9515	122.413	126.020
153	38	1982	38/82	2.8429	104.526	97.639
154	38	1984	38/84	3.2867	72.820	98.549

Price and Wage Observations, Worker Co-op Mills: Nos. 155–170

Obs	Mill No.	Year	ID	Real wage, w	Real log price, r	Real output price, p
155	38	1986	38/86	3.0479	74.461	95.083
156	4	1968	4/68	3.4137	91.178	124.002
157	4	1972	4/72	3.9151	91.890	123.133

Appendices

Obs	Mill No.	Year	ID	Real wage, w	Real log price, r	Real output price, p
158	4	1980	4/80	3.7398	122.413	126.020
159	4	1982	4/82	2.1832	104.526	97.639
160	4	1986	4/86	1.6284	74.461	95.083
161	5	1968	5/68	4.2971	86.379	124.002
162	5	1970	5/70	3.1711	79.196	97.445
163	31	1968	31/68	2.8058	95.977	124.002
164	31	1972	31/72	3.2260	98.682	123.133
165	31	1976	31/76	5.3714	149.684	143.755
166	31	1978	31/78	3.6374	133.190	166.027
167	31	1980	31/80	3.6753	107.415	126.020
168	31	1982	31/82	2.6817	107.988	97.639
169	31	1984	31/84	3.0881	83.176	98.549
170	31	1986	31/86	3.1489	84.879	95.083

References

Abbott, Michael, and Orley Ashenfelter. 1976. "Labour Supply, Commodity Demand, and the Allocation of Time." *Review of Economic Studies* 43(3): 389–411.

Agnew, T. H. 1917. "Report on the Health and Physical Condition of Male Munition Workers." Interim Report of the Health of Munition Workers Committee: Industrial Efficiency and Fatigue, 86–109. Cmd. 8511.

Anderson, Mary. 1936. "The Plight of Negro Domestic Labor." *Journal of Negro Education* 5(1): 66–72.

Anderson, Tosh. 2004. "'Overwork Robs Workers Health': Interpreting OSHA's General Duty Clause to Prohibit Long Work Hours." *New York City Law Review* 7: 85–160.

Anxo, Dominique, and Arne Bigsten. 1989. "Working Hours and Productivity in Swedish Manufacturing." *Scandinavian Journal of Economics* 91(3): 613–619.

Armstead, Robert. 2002. *Black Days, Black Dust: The Memories of an African-American Coal Miner*. Knoxville: University of Tennessee Press.

Ashenfelter, Orley. 1971. "The Effect of Unionization on Wages in the Public Sector: The Case of Fire Fighters." *Industrial and Labor Relations Review* 24(2): 637–650.

Ashenfelter, Orley, Kirk Doran, and Bruce Schaller. 2010. "A Shred of Credible Evidence on the Long-Run Elasticity of Labour Supply." *Economica* 77(308): 637–650.

Atack, Jeremy, and Fred Bateman. 1992. "How Long Was the Workday in 1880?" *Journal of Economic History* 52(1): 129–160.

References

Banks, Elizabeth L. 1894. *Campaigns of Curiosity: Journalistic Adventures of an American Girl in Late Victorian London*. Wisconsin Studies in Autobiography. Madison: University of Wisconsin Press.

Barnard, Catherine, Simon Deakin, and Richard Hobbs. 2003. "Opting-Out of the 48 Hour Week: Employer Necessity or Individual Choice? An Empirical Study of Article 18(1)(b) of the Working Time Directive in the U.K." *Industrial Law Journal* 32(4): 223–252.

Barger, Laura K., Brian E. Cade, Najib T. Ayas, John W. Cronin, Bernard Rosner, Frank Speizer, and Charles A. Czeisler. 2005. "Extended Work Shifts and the Risk of Motor Vehicle Crashes among Interns." *New England Journal of Medicine* 352: 125–134.

Becker, Gary S. 1977. "A Theory of the Production and Allocation of Effort." NBER Working Paper No. 184. National Bureau of Economic Research, Cambridge, MA. http://www.nber.org/papers/w0184

Berman, Katrina V. 1967. *Worker-Owned Plywood Companies*. Pullman: Washington State University Press.

Bick, Alexander, Bettine Brüggemann, and Nicola Fuchs-Schündeln. 2016. "Hours Worked in Europe and the U.S.: New Data, New Answers." IZA Discussion Paper No. 10179. Institute of Labor Economics, Bonn. http://ftp.iza.org/dp10179.pdf

Biddle, Jeff E., and Daniel S. Hamermesh. 1990. "Sleep and the Allocation of Time." *Journal of Political Economy* 98(5, Part 1): 9322–9343.

Bienefeld, Manfred A. 1972. *Working Hours in British Industry: An Economic History*. London School of Economics Research Monograph. Trowbridge: Redwood Press.

Bishop, Kelly, Bradley Heim, and Kata Mihaly. 2009. "Single Women's Labor Supply Elasticities: Trends and Policy Implications." *Industrial and Labor Relations Review* 63(1): 146–168.

Blau, Francine D., and Lawrence M. Kahn. 2007. "Changes in the Labor Supply Behavior of Married Women: 1980–2000." *Journal of Labor Economics* 25(3): 393–438.

Blaug, Mark. 1958. "The Classical Economists and the Factory Acts—A Re-Examination." *Quarterly Journal of Economics* 72(2): 211–226.

Blaug, Mark. 1997. *Economic Theory in Retrospect*, 5th ed. Cambridge: Cambridge University Press.

Boal, William M. 2016. "Work Intensity and Worker Safety in Early Twentieth-Century Coal Mining." Unpublished paper, Drake University. http://wmboal.com/research/Boal2017WorkIntensity.pdf

Boppart, Timo, and Per Krusell. 2016. "Labor Supply in the Past, Present, and Future: A Balanced-Growth Perspective." NBER Working Paper No. 22215. National Bureau of Economic Research, Cambridge, MA. http://www.nber.org/papers/w22215.pdf

Boulier, Bryan L. 1979. "Supply Decisions of Self-Employed Professionals: The Case of Dentists." *Southern Economic Journal* 45(3): 892–902.

Bourlès, Renaud, and Gilbert Cette. 2005. "A Comparison of Structural Productivity Levels in the Major Industrialised Countries." *OECD Economic Studies* 41:75–108.

Boushey, Heather. 2016. *Finding Time: The Economics of Work-Life Conflict.* Cambridge, MA: Harvard University Press.

Bowles, Samuel. 2004. *Microeconomics: Behavior, Institutions, and Evolution.* Russell Sage Foundation. Princeton, NJ: Princeton University Press.

Brachet, Timothy, Guy David, and Andrea M. Drechsler. 2012. "The Effect of Shift Structure on Performance." *American Economic Journal: Applied Economics* 4(2): 219–246.

Bradley, Ian Campbell. 1987. *Enlightened Entrepreneurs.* London: Weidenfeld and Nicolson.

Briggs, Asa. 1961. *Social Thought and Social Action; A Study of the Work of Seebohm Rowntree.* London: Longmans.

Brown, E. Henry Phelps. 1959. *The Growth of British Industrial Relations.* London: Macmillan.

Brown, E. Henry Phelps. 1983. *The Origins of Trade Union Power.* Oxford: Clarendon Press.

Browning, Martin, Angus Deaton, and Margaret Irish. 1985. "A Profitable Approach to Labor Supply and Commodity Demands over the Life-Cycle." *Econometrica* 53(3): 503–544.

Bry, Gerhard. 1959. "The Average Workweek as an Economic Indicator." NBER Occasional Paper No. 69. National Bureau of Economic Research, New York.

Bryan, Mark L. 2007. "Free to Choose? Differences in the Hours Determination of Constrained and Unconstrained Workers." *Oxford Economic Papers* 59(2): 226–252.

Buell, Philip, and Lester Breslow. 1960. "Mortality from Coronary Heart Disease in California Men Who Work Long Hours." *Journal of Chronic Diseases* 11(6): 615–626.

Burnett, John, ed. 1974. *Useful Toil: Autobiographies of Working People from the 1820s to the 1920s.* Harmondsworth, UK: Allen Lane.

References

Cahill, Marion. 1932. *Shorter Hours: A Study of the Movement since the Civil War.* New York: Columbia University Press.

Campbell, Janet M., and Lilian E. Wilson. 1917. "Inquiry into the Health of Women Engaged in Munition Factories." Interim Report of the Health of Munition Workers Committee: Industrial Efficiency and Fatigue, 110–121. Cmd. 8511.

Carrington, William J., Kristin McCue, and Brooks Pierce. 1996. "The Role of Employer/Employee Interactions in Labor Market Cycles: Evidence from the Self-employed." *Journal of Labor Economics* 14(4): 571–602.

Centers for Disease Control and Prevention. 2004. "Overtime and Extended Work Shifts: Recent Findings on Illnesses, Injuries, and Health Behaviors." National Institute for Occupational Safety and Health, US Department of Health and Human Services, Cincinnati, OH.

Clark, Gregory. 1987. "Productivity Growth without Technical Change in European Agriculture Before 1850." *Journal of Economic History* 47(2): 419–432.

Clark, Gregory. 1994. "Factory Discipline." *Journal of Economic History* 54(1): 128–163.

Coase, Ronald H. 1937. "The Nature of the Firm." *Economica* 4(16): 451–469.

Cobb, Charles W., and Paul H. Douglas. 1928. "A Theory of Production." *American Economic Review* 18(1): 761–785.

Collewet, Marion, and Jan Sauermann. 2017. "Working Hours and Productivity." IZA Discussion Paper No. 10722. Institute of Labor Ecoomics, Bonn. [To be published in in *Labour Economics*.] http://ftp.iza.org/dp10722.pdf

Conway, Sadie, Lisa Pompeii, Robert Roberts, Jack Follis, and David Gimeno. 2016. "Dose-Response Relation Between Work Hours and Cardiovascular Disease Risk." *Journal of Occupational & Environmental Medicine* 58(3): 221–226.

Coombes, Bert Lewis. 1939. *These Poor Hands: The Autobiography of a Miner Working in South Wales.* London: Victor Gollancz.

Cooper, C. L., M. J. Davidson, and P. Robinson. 1982. "Stress in the Police Service." *Journal of Occupational Medicine* 24(1): 30–36.

Costa, Dora L. 2000a. "The Wage and the Length of the Work Day: From the 1890s to 1991." *Journal of Labor Economics* 18(1): 156–181.

Costa, Dora L. 2000b. "Hours of Work and the Fair Labor Standards Act: A Study of Retail and Wholesale Trade, 1938–1950." *Industrial and Labor Relations Review* 53(4): 649–664.

References

Craig, Ben R., and John Pencavel. 1992. "The Behavior of Worker Cooperatives: The Plywood Companies of the Pacific Northwest." *American Economic Review* 82(5): 1083–1105.

Craig Ben R., and John Pencavel. 1995. "Participation and Productivity: A Comparison of Worker Cooperatives and Conventional Firms in the Plywood Industry." Brookings Papers on Economic Activity: Microeconomics, 121–60. Washington, DC.

Cross, Gary. 1989. *A Quest for Time: The Reduction of Work in Britain and France, 1840–1940.* Berkeley: University of California Press.

Curcio, Vincent. 2013. *Henry Ford.* Oxford: Oxford University Press.

Czeisier, Charles A. 2005. "Extended Work Shifts and the Risk of Motor Vehicle Crashes among Interns." *New England Journal of Medicine* 352(2): 125–134.

Dembe, Allard. E., J. B.Erickson, R. G.Delbos, and S. M. Banks. 2005. "The Impact of Overtime and Long Work Hours on Occupational Injuries and Illnesses; New Evidence from the United States." *Occupational and Environmental Medicine* 62(9): 588–597.

Dembe, Allard E., and Xiaoxi Yao. 2016. "Chronic Disease Risks from Exposure to Long-Hour Work Schedules over a 32-year Period." *Journal of Occupational and Environmental Medicine* 58(9): 861–867.

Denison, Edward F. 1962. *The Sources of Economic Growth in the United States and the Alternatives Before Us.* Committee for Economic Development, New York.

Devlin, Ciaran, and Alex Shirvani. 2014. "The Impact of the Working Time Regulations on the U.K. Labour Market: A Review of Evidence." UK Department for Business Innovation and Skills, London. https://www.gov.uk/government/uploads/system/uploads/attachment_data/file/389676/bis-14-12 87-the-impact-of-the-working-time-regulations-on-the-uk-labour-market-a-review-of-evidence.pdf

Dickinson, David L. 1999. "An Experimental Examination of Labor Supply and Work Intensities." *Journal of Labor Economics* 17(4): 638–670.

Earle, John S., and John Pencavel. 1990. "Hours of Work and Trade Unionism." *Journal of Labor Economics* 8(1, Part 2): S150–S174.

Edwards, George. 1922. *From Crow-Scaring to Westminster.* London: Labour Publishing. [Reprint by Larks Press, East Dereham, Norfolk, 2008.]

Ehrenberg, Ronald G. 1971. *Fringe Benefits and Overtime Behavior.* Lexington, MA: D.C. Heath.

References

Eichengreen, Barry. 1987. "The Impact of Late Nineteenth-Century Unions on Labor Earnings and Hours: Iowa in 1894." *Industrial and Labor Relations Review* 40(4): 501–515.

Ensor, Robert C. K. 1936. *England 1870–1914.* Oxford: Oxford University Press.

Eyer, Joseph. 1980. "Social Causes of Coronary Heart Disease." *Psychotherapy and Psychosomatics* 34: 75–87.

Fawcett, Henry. 1872. "The Regulation of the Hours of Labour by the State." In *Essays and Lectures on Social and Political Subjects,* ed. H. Fawcett and M. G. Fawcett. London: Macmillan.

Feldstein, Martin. 1967. "Specification of the Labour Input in the Aggregate Production Function." *Review of Economic Studies* 34(4): 375–386.

Feldstein, Martin. 1968. "Estimating the Supply Curve of Working Hours." *Oxford Economic Papers* 20(1): 74–80.

Fisher, Irving. 1909. "Report on National Vitality: Its Wastes and Conservation." Bulletin of the Committee of One Hundred on National Health. National Conservation Commission, Washington, DC.

Foster, Andrew D., and Rosenzweig, M. R. 1994. "A Test for Moral Hazard in the Labor Market: Contractual Arrangements, Effort, and Health." *Review of Economics and Statistics* 76(2): 213–227.

Frandsen, Brigham R. 2016. "The Effects of Collective Bargaining Rights on Public Employee Compensation: Evidence from Teachers, Firefighters, and Police." *Industrial and Labor Relations Review* 69(1): 84–112.

Fried, Jason, and David Hansson. 2010. *ReWork.* New York: Crown.

Fritz, C., S. Sonnentag, P. Spector, and J. McInroe. 2010. "The Weekend Matters: Relationships Between Stress Recovery and Effective Experiences." *Journal of Organizational Behavior* 31: 1137–1162.

Garnero, Andrea, Stephan Kampelmann, and François Rycx. 2014. "Part-Time Work, Wages, and Productivity: Evidence from Belgian Matched Panel Data." *Industrial and Labor Relations Review* 67(3): 926–954.

Golden, Lonnie, and Tesfayi Gebreselassie. 2007. "Overemployment Mismatches: The Preference for Fewer Work Hours." *Monthly Labor Review* 4: 18–37.

Golden, Lonnie, and Barbara Wiens-Tuers. 2008. "Overtime Work and Wellbeing at Home." *Review of Social Economy* 66(1): 25–49.

Goldin, Claudia 1988. "Maximum Hours Legislation and Female Employment: A Reassessment." *Journal of Political Economy* 96(1): 189–205.

References

Goldmark, Josephine. 1912. *Fatigue and Efficiency: A Study in Industry*. New York: Russell Sage Foundation.

Gompertz, Benjamin. 1825. "On the Nature of the Function Expressive of the Law of Human Mortality, and on a New Mode of Determining the Value of Life Contingencies." *Philisophical Transactions of the Royal Society of London* 115: 513–583.

Grandjean, Etienne. 1969. *Fitting the Task to the Man: An Ergonomic Approach*. London: Taylor and Francis.

Green, Francis. 2001. "It's Been a Hard Day's Night: The Concentration and Intensification of Work in late Twentieth-Century Britain." *British Journal of Industrial Relations* 39(1): 53–80.

Griliches, Zvi 1967. "Production Functions in Manufacturing: Some Preliminary Results." In *The Theory and Empirical Analysis of Production*, ed. Murray Brown, 275–322. Studies in Income and Wealth, vol. 31. National Bureau of Economic Research. New York: Columbia University Press.

Hamermesh, Daniel S., and Elena Stancanelli. 2014. "Long Workweeks and Strange Hours." IZA Discussion Paper No. 8423. Institute of Labor Economics, Bonn. http://ftp.iza.org/dp8423.pdf

Hart, Robert A. 1987. *Working Time and Employment*. Boston: Allen & Unwin.

Hart, Robert A., and Peter G. McGregor. 1988. "The Returns to Labour Services in West German Manufacturing Industry." *European Economic Review* 32: 947–963.

Health and Safety Laboratory. 2003. "Working Long Hours." Sheffield University. http://www.hse.gov.uk/research/hsl_pdf/2003/hsl03-02.pdf

Health of Munition Workers Committee (HMWC). 1919. "Final Report." Cmd. 9065., London.

Heim, Bradley T. 2007. "The Incredible Shrinking Elasticities: Married Female Labor Supply, 1978–2002." *Journal of Human Resources* 42(4): 881–918.

Hicks, John R. 1963. *The Theory of Wages*, 2nd ed. New York: Macmillan. [First edition, 1932.]

Hobsbawm, E. J. 1952. "The Machine Breakers." *Past and Present* 1: 57–70.

Hobsbawm, E. J. 1974. "British Gas Workers 1873–1914." Chapter 9 of his *Labouring Men: Studies in the History of Labour*, 158–178. New York: Basic Books.

Huberman, Michael, and Chris Minns. 2007. "The Times They Are Not Changin: Days and Hours of Work in Old and New Worlds, 1870–2000." *Explorations in Economic History* 44: 538–567.

References

Hunnicutt, Benjamin K. 1984. "The End of Shorter Hours." *Labor History* 25(3): 373–404.

Hunnicutt, Benjamin K. 1996. *Kellogg's Six-Hour Day*. Philadelphia: Temple University Press.

Hutchins, B. L., and Amy Harrison. 1966. *A History of Factory Legislation*, 3rd. ed. New York: Augustus M. Kelly. [First edition, 1903.]

Industrial Health Research Board (IHRB). 1942. "Hours of Work, Lost Time and Labour Wastage." Emergency Report No. 2, Medical Research Council, London.

Ineson, Antonia, and Deborah Thom. 1985. "T.N.T. Poisoning and the Employment of Women Workers in the First World War." In *The Social History of Occupational Health*, ed. Paul Weindling, 89–107. Beckenham, Kent: Croom Helm.

Iwasaki, Kenji, Masaya Takahashi, and Akinori Nakata. 2006. "Health Problems due to Long Working Hours in Japan." *Industrial Health* 44: 537–540.

Jansen, N., I. Kant, and P. A. van den Brandt. 2002. "Need for Recovery in the Working Population: Description and Associations with Fatigue and Psychological Distress." *International Journal of Behavioral Medicine* 94(4): 322–340.

Jones, Ethel B. 1975. "State Legislation and Hours of Work in Manufacturing." *Southern Economic Journal* 41: 602–612.

Kaufman, Roger T. 1984. "On Wage Stickiness in Britain's Competitive Sector." *British Journal of Industrial Relations* 22(1): 101–112.

Keane, Michael P. 2011. " Labor Supply and Taxes: A Survey." *Journal of Economic Literature* 59(4): 961–1075.

Killingsworth. Mark R. 1983. *Labor Supply*. Cambridge: Cambridge University Press.

Kivimaki, Mika, Markus Jokela, et al. 2015. "Long Working Hours and Risk of Coronary Heart Disease and Stroke." *The Lancet* 386: 1739–1746.

Kodz, Jenny, Sara Davis, David Lain, Marie Strebler, Jo Rick, Peter Bates, John Cummings, and Nigel Meager. 2003. "Working Long Hours: A Review of the Evidence." Employment Relations Research Series No. 16. Institute for Employment Studies, Department of Trade and Industry, London. http://www.employment-studies.co.uk/system/files/resources/files/errs16_main.pdf

Kossoris, Max D. 1944. "Studies of the Effects of Long Working Hours." *Monthly Labor Review* 58(6): 1131–1144.

Kossoris, Max D., and Reinfried F. Kohler. 1948. "Hours of Work and Output." Bureau of Labor Statistics, Bulletin No. 917, US Department of Labor, Washington, DC.

Kosters, Marvin H. 1966. "Income and Substitution Effects in a Family Labor Supply Model." Report P-3339, The Rand Corporation, Santa Monica, California.

Kuhn, Peter, and Fernando Lozano. 2008. "The Expanding Workweek? Understanding Trends in Long Work Hours among U.S. Men, 1979–2006." *Journal of Labor Economics* 26(2): 311–343.

Künn-Nelen, AnneMarie, Andres de Grip, and Didier Fouarge. 2013. "Is Part-time Employment Beneficial for Firm Productivity?" *Industrial and Labor Relations Review* 66(5): 1172–1191.

Landrigan, Christopher P., Laura K. Barger, Brian E. Cade, Najib T. Ayas, and Charles A. Czeisler. 2006. "Interns' Compliance with Accreditation Council for Graduate Medical Education Work-Hour Limits." *Journal of the American Medical Association* 296(9): 1063–1070.

Landrigan Christopher P., J. M. Rothschild, J. W. Cronin, R. Kaushal, E. Burdick, J. T. Katz, C. M. Lilly, P. H. Stone, S. W. Lockley, D. W. Bates, and C. A.Czeisler. 2004. "Effect of Reducing Interns' Work Hours on Serious Medical Errors in Intensive Care Units." *New England Journal of Medicine* 351: 1838–1848.

Laslett, Peter. 1984. *The World We Have Lost*, 3rd ed. New York: Charles Scribner's Sons.

Lazear, Edward P. 2000. "Performance Pay and Productivity." *American Economic Review* 90(5): 1346–1361.

Lebergott, Stanley. 1964. *Manpower in Economic Growth: The American Record Since 1800*. New York: McGraw-Hill.

Lee, Jungmin, and Yong-Kwan Lee. 2016. "Can Working Hour Reduction Save Workers?" *Labour Economics* 40: 25–36.

Lester, Richard A. 1946. *Economics of Labor*. New York: Macmillan.

Lethbridge, Lucy. 2013. *Servants: A Downstairs View of Twentieth Century Britain*. London: Bloomsbury.

Lewis, H. Gregg. 1957. "Hours of Work and Hours of Leisure." In *Proceedings of the Ninth Annual Meeting of the Industrial Relations Research Association*, ed. L. Reed Tripp, 196–206. Madison, WI: Industrial Relations Research Association.

Lewis, H. Gregg. 1963. *Unionism and Relative Wages in the United States: An Empirical Inquiry*. Chicago: University of Chicago Press.

Lewis, H. Gregg. 1969. "Employer Interests in Employee Hours of Work" ["Interes del empleador en las horas de trabajo del empleado"]. *Cuadernos de Economia* 6(18): 38–54.

Lewis, H. Gregg. 1986. *Union Relative Wage Effects: A Survey.* Chicago: University of Chicago Press.

Lin, Chung-cheng. 2003. "A Backward-Bending Labor Supply Curve Without an Income Effect." *Oxford Economic Papers* 55(2): 336–343.

Loveday, Thomas. 1917. "The Causes and Conditions of Lost Time." Industrial Efficiency and Fatigue, Health of Munition Workers Committee, Interim Report, 41–67, Cd. 8511.

Lowe, Rodney. 1982. "Hours of Labour: Negotiating Industrial Legislation in Britain, 1919–39." *Economic History Review* 35(2): 254–271.

MacRae-Gibson, J. H. 1922. *The Whitley System in the Civil Service.* Westminster: Fabian Society.

MaCurdy, Thomas E. 1981. "An Empirical Model of Labor Supply in a Life-Cycle Setting." *Journal of Political Economy* 89(6): 1059–1085.

Marmot, Michael, and Eric Brunner. 2005. "Cohort Profile: The Whitehall II Study." *International Journal of Epidemiology* 34(2): 251–256.

Marshall, Alfred. 1920. *Principles of Economics*, 8th ed. London: Macmillan.

Marwick, Arthur. 1970. *The Deluge: British Society and the First World War.* New York: W.W. Norton.

Mather, William. 1894. *The Forty-Eight Hours Week: A Year's Experiment and Its Results at the Salford Iron Works, Manchester.*" Manchester: Guardian Printing Works.

Matthews, Robin C. O., Charles. H. Feinstein, and John Odling-Smee. 1982. *British Economic Growth 1856–1972.* Stanford, CA: Stanford University Press.

McCartt, Anne T., John W. Rohrbaugh, Mark C. Hammer, and Sandra Z. Fuller. 2000. "Factors Associated with Falling Asleep at the Wheel Among Long-Distance Truck Drivers." *Accident & Prevention* 32: 493–504.

McIvor, A. J. 1987. "Employers, the Government, and Industrial Fatigue in Britain, 1890–1918." *British Journal of Industrial Medicine* 44: 724–732.

Merish, Lori. 2017. *Archives of Labor: Working Class Women and Literary Culture in the Antebellum United States.* Durham, NC: Duke University Press.

Meyer III, Stephen. 1981. *The Five Dollar Day: Labor Management and Social Control in the Ford Motor Company 1908–1921.* Albany: State University of New York Press.

References

Millis, Harry A., and Royal E. Montgomery. 1938. *Labor"s Progress and Some Basic Labor Problems*. New York: McGraw-Hill.

Mincer, Jacob. 1974. *Schooling, Experience and Earnings*. National Bureau of Economic Research. New York: Columbia University Press.

Mitchell, B. R. 1988. *British Historical Statistics*. Cambridge: Cambridge University Press.

National Industrial Conference Board. 1917. "Analysis of British Wartime Reports on Hours of Work as Related to Output and Fatigue." Report No. 2, November, Boston, MA.

Nyland, Chris. 1989. *Reduced Worktime and the Management of Production*. Cambridge: Cambridge University Press.

Oi. Walter Y. 1973. "An Essay on Workmen"s Compensation and Industrial Safety." Supplemental Studies for the National Commission on State Workmen's Compensation Laws, 41–106. Washington, DC.

Organization of Economic Cooperation and Development. 2016. *Employment Outlook 2016*. Paris: OECD Publishing.

Owen, Robert. 1816. "Report from the Select Committee on the State of the Children Employed in the Manufactories of the United Kingdom." Minutes of Evidence, April 25–June 18. British Sessional Papers, vol. 3.

Paarsch, Harry J., and Bruce S. Shearer. 1999. "The Response of Worker Effort to Piece Rates: Evidence from the British Columbia Tree-Planting Industry." *Journal of Human Resources* 34(4): 643–667.

Parker, Simon C., Yacine Belghitar, and Tim Barmby. 2005. "Wage Uncertainty and the Labour Supply of Self-Employed Workers." Conference Papers. *Economic Journal* 115(502): C190–C207.

Pencavel, John. 1977. "Work Effort, On-the-Job Screening, and Alternative Methods of Remuneration." *Research in Labor Economics* 1: 225–258.

Pencavel, John. 1986. "Labor Supply of Men." In *Handbook of Labor Economics*, vol. 1, ed.. Orley Ashenfelter and Richard Layard, 3–102. Amsterdam: Elsevier Science.

Pencavel, John. 2001. *Worker Participation: Lessons from the Worker Co-ops of the Pacific Northwest*. New York: Russell Sage Foundation.

Pencavel, John. 2004. "The Surprising Retreat of Union Britain." In *Seeking a Premier Economy: The Economic Effects of British Economic Reforms, 1980-2000*, ed. David Card, Richard Blundell, and Richard B. Freeman, 181–232. National Bureau of Economic Research. Chicago: University of Chicago Press.

References

Pencavel, John. 2015a. "The Labor Supply of Self-Employed Workers: The Choice of Working Hours in Worker Co-ops." *Journal of Comparative Economics* 43(3): 677–689.

Pencavel, John. 2015b. "The Productivity of Working Hours." *Economic Journal* 125(589): 2052–2076.

Pencavel, John. 2016a. "Whose Preferences Are Revealed in Hours of Work?" *Economic Inquiry* 54(1): 9–24.

Pencavel, John. 2016b. "Recovery from Work and the Productivity of Working Hours." *Economica* 83(332): 545–563.

Pencavel, John, and Ben Craig. 1994. "The Empirical Performance of Orthodox Models of the Firm: Conventional Firms and Worker Cooperatives." *Journal of Political Economy* 102(4): 718–744.

Proctor, Susan P., Roberta F. White, Thomas G. Robins, Diana Echeverria, and Adrian Z. Rocskay. 1996. "Effect of Overtime Work on Cognitive Function in Automotive Workers." *Scandinavian Journal of Work, Environment & Health* 22(2): 124–132.

Rae, John. 1894. *Eight Hours for Work.* London: Macmillan.

Ray, Rebecca, and John Schmitt. 2007. "No-vacation Nation USA—A Comparison of Leave and Holiday in OECD Countries." European Economic and Employment Policy Brief No. 3, ISSN 1782–2165, Brussels.

Reder, Melvin W. 1999. *Economics: The Culture of a Controversial Science.* Chicago: University of Chicago Press.

Reynolds, Lloyd G. 1954. *Labor Economics and Labor Relations*, 2nd ed. New York: Prentice-Hall.

Ricci, Judith A., Elsbeth Chee, Amy L. Lorandeau, and Jan Berger. 2007. "Fatigue in the U.S. Workforce: Prevalence and Implications for Lost Productive Work Time." *Journal of Occupational and Environmental Medicine* 49(1): 1–10.

Robbins, Lionel. 1978. *The Theory of Economic Policy*, 2nd ed. Philadelphia: Porcupine Press.

Rogers, Ann E., Wei-Ting Hwang, Linda D. Scott, Linda H. Aiken, and David F. Dinges. 2004. "The Working Hours of Hospital Staff Nurses and Patient Safety." *Health Affairs* 23(4): 202–212.

Rosen, Sherwin. 1969. "On the Interindustry Wage and Hours Structure." *Journal of Political Economy* 77(2): 249–273.

Rosen, Sherwin. 1974. "Hedonic Prices and Implicit Markets: Product Differentiation in Pure Competition." *Journal of Political Economy* 82(1): 34–55.

Rubin, Marcus, and Ray Richardson. 1997. *The Microeconomics of the Shorter Working Week*. Aldershot: Avebury.

Russek, H. I., and B. L. Zohman. 1958. "Relative Significance of Heredity, Diet, and Occupational Stress in Coronary Heart Disease of Young Adults; Based on an Analysis of 100 Patients Between the Ages of 25 and 40 Years and a Similar Group of 100 Normal Control Subjects." *American Journal of the Medical Sciences* 235(3): 266–277.

Salanié, Bernard. 2005 *The Economics of Contracts*, 2nd ed. Cambridge, MA: MIT Press.

Schneiderman, Rose. 1905. "A Cap-Maker's Story." *The Independent* 58(2943): 935–938.

Schulte, Brigid. 2014. *Overwhelmed: Work, Love, and Play When No-one Has the Time*. New York: Picador.

Schumpeter, Joseph A. 1954. *History of Economic Analysis*. New York: Oxford University Press.

Scotland, Nigel. 1997. "Methodism and the English Labour Movement 1800–1906." *Anvil: A Journal of Theology and Mission* 14(1): 36–48.

Sharpless, Rebecca. 2010. *Cooking in Other Women's Kitchens: Domestic Workers in the South, 1865-1960*. Chapel Hill: University of North Carolina Press.

Shepard, Edward, and Thomas Clifton. 2000. "Are Longer Hours Reducing Productivity in Manufacturing?" *International Journal of Manpower* 21(7): 540–552.

Shetty, Kanaka D., and Jayanta Bhattacharya. 2007. "Changes in Mortality Associated with Residency Work-Hour Regulations." *Annals of Internal Medicine* 147(2): 73–80.

Shiells, Martha. 1990. "Collective Choice of Working Conditions: Hours in British and U.S. Iron and Steel, 1890–1923." *Journal of Economic History* 50(2): 379–392.

Simon, Herbert A. 1951. "A Formal Theory of the Employment Relationship." *Econometrica* 19(3): 293–305.

Smith, Adam. 1776. *An Inquiry into the Nature and Causes of the Wealth of Nations*. London: W. Strahan and T. Cadell.

Smuts, Robert W. 1953. *European Impressions of the American Worker*. New York: King's Crown Press, Columbia University Press.

Sonnentag, S., and F. Zijlstra. 2006. "Job Characteristics and Off-job Activities as Predictors of Need for Recovery, Well-being, and Fatigue." *Journal of Applied Psychology* 91(2): 330–350.

References

Sorenson, Lloyd R. 1952. "Some Classical Economists, Laissez Faire and the Factory Acts." *Journal of Economic History* 12(3): 247–262.

Spurgeon, Anne, and J. Malcolm Harrington. 1989. "Work Performance and Health of Junior Hospital Doctors: A Review of the Literature." *Work & Stress* 3(2): 117–128.

Spurgeon, Anne, J. Malcolm Harrington, and Cary L. Cooper. 1997. "Health and Safety Problems Associated with Long Working Hours: A Review of the Current Position." *Occupational and Environmental Medicine* 54: 367–375.

Stewart, Mark B., and Joanna K. Swaffield. 1997. "Constraints on the Desired Hours of Work of British Men." *Economic Journal* 107(441): 520–535.

Stigler, George J. 1941. *Production and Distribution Theories: The Formative Period*. New York: Macmillan.

Sundstrom, William A. 2006. "Hours and Working Conditions." In *Historical Statistics of the United States Millenial Edition*, vol. 2, ed. S. Carter, S. S. Gartner, M. R. Haines, A. L. Olmstead, R. Sutch, and G. Wright, 301–335. New York: Cambridge University Press.

Terkel, Studs. 1975. *Working*. New York: Avon.

Treble, John G. 2001. "Productivity and Effort: The Labour Supply Decisions of Late Victorian Coalminers." *Journal of Economic History* 61(2): 414–438.

Treble, John G. 2003. "Intertemporal Substitution of Effort: Some Empirical Evidence." *Economica* 70(280): 579–595.

Trejo, Stephen J. 1993. "Overtime Pay, Overtime Hours, and Labor Unions." *Journal of Labor Economics* 11(2): 253–278.

Troy, Leo, and Neil Sheflin. 1985. "Union Sourcebook: Membership, Structure, Finance." Directory, Industrial Relations Data Information Services, West Orange, NJ.

Uehata, Tetsunojo. 1991. "Long Working Hours and Occupational Stress-related Cardiovascular Attacks Among Middle-aged Workers in Japan." *Journal of Human Ergology* 20(2): 147–153.

US Department of Labor, Bureau of Labor Statistics. 1917a. "British Munition Factories." Reprints of the Memoranda of the British Health of Munition Workers Committee, Bulletin No. 221, Washington, DC.

US Department of Labor, Bureau of Labor Statistics. 1917b. "Welfare Work in British Munition Factories." Reprints of the Memoranda of the British Health of Munition Workers Committee, Bulletin No. 222, Washington, DC.

US Department of Labor, Bureau of Labor Statistics. 1917c. "Employment of Women and Juveniles in Great Britain during the War." Reprints of the Memoranda of the British Health of Munition Workers Committee, Bulletin No. 223,Washington, DC.

US Department of Labor, Bureau of Labor Statistics. 1917d. "Industrial Efficiency and Fatigue in British Munition Factories." Reprints of the Memoranda of the British Health of Munition Workers Committee, Bulletin No. 230, Washington, DC.

US Department of Labor, Bureau of Labor Statistics. 1919. "Industrial Health and Efficiency." Final Report of the British Health of Munition Workers Committee, Bulletin No. 249, Washington, DC.

Van Soest, Arthur, Isolde Woittiez, and Arie Kapteyn. 1990. "Labor Supply, Income Taxes, and Hours Restrictions in the Netherlands." *Journal of Human Resources* 25(3) 517–558.

Vernon, Horace M. 1917. "Further Statistical Information concerning Output in Relation to Hours of Work with Special Reference to the Influence of Sunday Labour." Health of Munition Workers Committee Memorandum No. 18, Cd. 8628.

Vernon, Horace M. 1921. *Industrial Fatigue and Efficiency.* London: Routledge.

Vila, Bryan. 2006. "Impact of Long Hours on Police Officers and the Communities They Serve." *American Journal of Industrial Medicine* 49(11): 972–980.

Virtanen, Marianna, Jane E. Ferrie, Archana Singh-Manoux, Martin Shipley, Jussi Vahtera, Michael G. Marmot, and Mika Kivimaki. 2010. "Overtime Work and Incident Heart Disease: The Whitehall II Prospective Cohort Study." *European Heart Journal* 31(14): 1737–1744.

Virtanen, Marianna, Archana Singh-Manoux, Jane E. Ferrie, David Gimeno, Michael G. Marmot, Marko Elovainio, Markus Jokela, Jussi Vahtera, and Mika Kivimä. 2009. "Long Working Hours and Cognitive Function: The Whitehall II Study." *American Journal of Epidemiology* 169(5): 596–605.

Webb, Sidney, and Harold Cox. 1891. *The Eight Hours Day.* London: Walter Scott.

Whaples, Robert. 1990. "Winning the Eight-Hour Day, 1909–1919." *Journal of Economic History* 50(2): 393–406.

White, Michael. 1987. *Working Hours: Assessing the Potential for Reduction.* Geneva: International Labour Office.

White, Michael, and Alex Bryson. 2016. "Unions and the Economic Basis of Attitudes." IZA Discussion Paper No. 9876, Institute of Labor Economics, Bonn.

White, Michael, Stephen Hill, Patrick McGovern, Colin Mills, and Deborah Smeaton. 2003. "High-performance Management Practices, Working Hours and Work-Life Balance." *British Journal of Industrial Relations* 41(2): 175–195.

Williams, Iolo A. 1931. *The Firm of Cadbury 1831–1931*. London: Constable.

Wilson, Harold. 1964. *The Relevance of British Socialism*. London: Weidenfeld and Nicolson.

Wohl, Anthony S. 1983. *Endangered Lives: Public Health in Victorian Britain*. London: J. M.Dent and Sons.

Wolman, Leo. 1924. *The Growth of American Trade Unions 1880–1923*. New York: National Bureau of Economic Research.

Wright, Gavin. 2010. "The Industrious Revolution in America." In *The Birth of Modern Europe: Culture and Economy 1400–1800: Essays in Honor of Jan De Vries*, ed. Laura Cruz and Joel Mokyr, 301–335. Leiden: Brill.

Index

Figures, notes, and tables are indicated by *f*, n, and *t* following the page number.

Abbé, Ernst, 26, 70
Abbott, Michael, 162
absenteeism, 105n17, 109–111, 109*t*
accidents, job-related, 148–149.
　See also health issues of workers
Accreditation Council on Medical Education, 142
Acuna, Roberto, 14
Agnew, T. H., 139
agricultural work
　post-Industrial Revolution, 14–16, 19
　pre-Industrial Revolution, 13–14
　work effort and, 56–57
Amalgamated Society of Engineers, 25
American Federation of Labor (AFL), 46–47
Anderson, Mary, 14
Anderson, Tosh, 141
Anxo, Dominique, 62n3
Armstead, Robert, 149

Ashenfelter, Orley, 23, 162, 178
Asquith, H. H., 69
Atack, Jeremy, 42
average product of hours
　Cobb-Douglas production function and, 61, 61*f*
　Gompertz production function and, 81, 82*f*

Banks, Elizabeth, 15
Bateman, Fred, 42
Berman, Katrina, 173n9, 176, 177
Bhattacharya, Jayanta, 142
Bick, Alexander, 10
Biddle, Jeff E., 184
Bienefeld, Manfred A., 19, 21, 24, 39, 153, 179
Bigsten, Arne, 62n3
Blaug, Mark, 32, 50
Boal, William, 138n1
Boppart, Timo, 42n17
Boulier, Bryan L., 178–179

243

Index

Bourlès, Renaud, 137
Bowles, Samuel, 54
Brachet, Timothy, 136
Breslow, Lester, 144
Britain. *See also* case studies of production functions
   Act to Limit the Hours of Labour of Young Persons and Females in Factories (1847), 31
   agricultural and domestic work, pre- and post-Industrial Revolution, 13–16, 19
   average weekly hours of work in, 7–12, 9*t*, 181
   Department for Business Innovation and Skills, 35
   Factory Acts (1802), 30–31, 33
   Health and Morals of Apprentices Act (1802), 31
   historical reduction of work hours in, 23–27, 29–35, 42–48, 48n19, 182–184
   majority of workers as employees in, 153
   Shop Hours Regulation Act (1886), 34
   Trade Disputes Act (1906), 45
   wage-earnings relationship in, 167–168
   Whitley Council for the Administrative and Legal Departments of the Civil Service (1919), 34
   working hours in, overview, 3–4
Brown, E. Henry Phelps, 21, 45
Brown, Lewis J., 28
Browning, Martin, 159
Brunner, Eric, 143

Bry, Gerhard, 159
Buell, Philip, 144
Bureau of Labor Statistics (US), 150

Cadbury, George, 29, 189
Cadbury, Richard, 29, 189
Cahill, Marion, 20, 28, 35, 48–49
California, legislation on work hours in, 36
calories, work effort and, 56
Cambridge University, 188–189
Campbell, Janet M., 139
cardiovascular disease, 144–145
case studies of production functions, 68–132. *See also individual case studies*
   Craig (case study), 111–125
   data collection variation across case studies, 103–105, 125–126
   hours and output analysis, 126–132, 129–131*f*
   Industrial Health Research Board (case study), 95–104
   Kossoris and Kohler (case study), 104–111
   overtime work and, 136–137
   part-time vs. full-time workers, 133–136
   Vernon (case study), 68–95
   CD-1 (Cobb-Douglas conventional specification)
   Cobb-Douglas production function, components of, 57–58
   Cobb-Douglas production function, elasticity of, 65
   Craig's observations and, 123

days of work and output, 86,
  86n10, 89, 91t
 defined, 68–72
 hours-output relationship, IHRB
  on, 96–102, 99t, 102t
 hours-output relationship, Vernon
  on, 74–80, 77t, 78–79f, 81t
 labor productivity and, 134
 marginal product of working
  hours, 61, 61f, 78, 79f
 overtime work and, 136–137
 proportional changes in hours vs.
  output, 95, 100–101, 102t
CD-2 (Cobb-Douglas augmented
  specification)
 days of work and output, 86,
  86n10, 89, 91t
 defined, 62, 185
 hours-output relationship, IHRB
  on, 97–104, 99t, 100–101f, 103f
 hours-output relationship, Vernon
  on, 74, 76–78, 77t, 78–80f
 labor productivity and, 134
 marginal product of working
  hours, 61, 61f, 78, 79f
Census of Manufactures, 36
Centers for Disease Control and
  Prevention, 141
Cette, Gilbert, 137
children as workers
 historical reduction in work hours
  of, 16, 26–27, 31–35, 33n14
 munition factory workers during
  First World War, 69, 71, 73. See
  also Vernon case study
 statutory legislation on, 30–31
civil servants (British), health studies
  of, 142–144

Clark, Cyrus, 29, 189
Clark, Gregory, 16–17, 56
Clark, James, 29, 189
classical school of economic thought
 historical reduction in work hours
  and, 49–50
 on statutory legislation of
  work, 32
Clifton, Thomas, 137
co-operative ownership. See also
  Craig case study
 conventional ownership
  vs., 122–125, 123t, 190
 self-employed workers and hours-
  earnings association, 172–180,
  174–177nn10–16, 176t, 178f
coal miners
 early formalization of work hours
  and, 16
 health issues of, 138n1
 piece-rate pay of, 55–56
Cobb, Charles. See also Cobb-Douglas
  production function
 Cobb-Douglas production
  function of, 57–62
 research of, 51
Cobb-Douglas production function.
  See also CD-1; CD-2
 augmented specification (CD-2),
  defined, 62, 185
 components of, 57–58
 conventional specification
  (CD-1), defined, 58–62
 as elasticity of output with respect
  to hours, 65
cognitive function of workers,
  143–144
Collewet, Marion, 135

Index

compensation. *See* earnings and wages
Connecticut, legislation on work hours in, 35
constant returns to hours
   Craig's observations on, 118, 118*f*
   defined, 51
   hours-output relationship, IHRB on, 95, 98–103, 100–101*f*, 102*t*
   unit returns to hours vs., 52–53
construction, reduction of work hours in, 20–21
consumption of commodities, 157, 162, 162n6. *See also* hours-earnings relationship
Conway, Sadie, 145–146
Coombes, Bert, 148–149
Costa, Dora, 162–164, 164–165*f*
Craig case study, 111–125
   conventional ownership vs. co-operative mills, 122–125, 123*t*, 190
   data collection methods, 111–116, 113–114*t*
   data used in, 211–225, 213–215*t*, 218–221*t*, 224*t*
   diminishing returns to hours conclusion for, 126–132, 131*f*
   hours and output, 116–118, 117*t*, 118–119*f*
   lump of labor proposition, 119–122, 120*f*, 121n21

Da Vinci, Philip, 141
demand
   demand for hours per worker, 122, 123*t*, 153–156. *See also* hours-earnings relationship

"labor supply" as synonym for "hours," 164–169
reduction in hours vs. rise in pay, 39–42, 41*f*
Dembe, Allard, 146, 148
Denison, Edward, 167
dentists (nonsalaried), hours-earnings relationship for, 178–179
Devlin, Ciaran, 34–35
domestic work, 13–16, 19
Douglas, Paul
   Cobb-Douglas production function of, 57–62
   research of, 51. *See also* CD-1; CD-2; Cobb-Douglas production function

earnings and wages. *See also* hours-earnings relationship
   historical reduction of work hours and, 16–18, 33n14
   overtime pay, 16, 49, 136–137, 153n2
   standard of living and, 1, 38–39
   Sunday work and additional pay, 107
   work effort and, 4, 54–57
Edwards, George, 46
effective hours
   defined, 4, 53
   nominal hours vs., 54–57, 185–186
Ehrenberg, Ronald G., 153n2
Eichengreen, Barry, 21n7
elasticity of output, 50, 60–62, 60n2. *See also* hours-output relationship

employees. *See also* hours-earnings relationship; hours-output relationship
   average weekly hours of, 5–12, 6t, 168, 181
   effective hours of, 4, 53–57, 185–186
   individual workers' knowledge of job requirements, 189–190
   majority of workers as, 153
   plant's production function as conduct of employees, 66–67
   reducing unemployment by reducing hours of work for, 18
   supply of worker hours by. *See* hours-earnings relationship
   worker-hours, defined, 5
"Employer Interests in Employee Hours of Work" (Lewis), 160
employers. *See also* co-operative ownership
   demand for worker hours by. *See* hours-earnings relationship
   history of reduction in work hours, 23–30
   plant's production function as conduct of employees, 66–67
   production function and importance to, 1–2, 187–191
Ensor, Robert C. K., 45
estimates of production functions. *See* case studies of production functions
European Union, Working Time Directive (1998) of, 34–35
Evett, Paul, 44
Eyer, Joseph, 144

"factory discipline," 16–17, 17n3
fatigue. *See also* health issues of workers
   as contemporary issue, 140–142
   performance and, 65
   recovery from work and, 88–94, 90–91t, 93–94t
Fawcett, Henry, 33, 188–189
Federation of Engineering Employers, 26
Feldstein, Martin, 59, 161
Ford, Henry, 27–29, 189
Foster, Andrew D., 56
Frandsen, Brigham R., 23
full-time vs. part-time work, 133–136

Garnero, Andrea, 135
gas workers (Britain), reduction of work hours for, 20
gender issues
   absenteeism and, 110–111
   health issues and, 139, 147
   historical reduction in work hours and, 16, 42, 44
   hours-earnings relationship and, 163–164, 164–165f, 171–172
   statutory legislation on work and, 31–33, 35–37
   women as munition factory workers, 69, 69n1, 71–72, 95–96
   work effort and, 54
Germany, average weekly hours of work (1979 vs. 2015), 9, 10t
Goldin, Claudia, 37
Gompers, Samuel, 18

Index

Gompertz, Benjamin.
  See GZ-1; GZ-2
Grandjean, Etienne, 65
Great Britain. *See* Britain
Great Recession, 6n2
Green, Francis, 54
Griliches, Zvi, 59
GZ-1 (Gompertz conventional specification)
  days of work and output, 86, 86n10, 91t
  defined, 62–67
  hours-output relationship, IHRB on, 98, 99t
  hours-output relationship, Vernon on, 74, 76, 77t
GZ-2 (Gompertz augmented specification)
  days of work and output, 86, 86n10, 89, 91t
  defined, 63–67, 185
  hours-output relationship, IHRB on, 98–104, 99t, 100–101f
  hours-output relationship, Vernon on, 74, 76, 77t, 78f, 80–81, 82f, 84
  marginal product of working hours, 81–88, 82f

Hamermesh, Daniel S., 184
Hart, Robert A., 136
health and Safety Laboratory, 141
health issues of workers, 138–150
  calories consumed and work effort, 56
  cardiovascular disease and, 144–145
  cognitive function and, 143–144
  fatigue and, 65, 88–94, 90–91t, 93–94t, 140–142
  household well-being and, 149–150
  importance of understanding implications of, 186–187
  injuries and accidents resulting from, 148–149
  Kossoris and Kohler on, 105n18, 109–111, 109t
  of munition workers, 138–140
  in nationally representative populations, 144–147
  of textile workers vs. agricultural workers, 31n12
Health of Munition Workers Committee (HMWC), 69–72, 138–139, 193.
  See also Vernon case study
hedonics model, 41–42
Hicks, John R., 22, 49, 54, 188
history of working hours, 13–51
  of agricultural and domestic work, 13–16, 19
  in Britain vs. United States, 42–48.
  See also Britain; United States
  employers' influence on, 25–30
  formalization of working hours, 15–17
  interdependence of factors on, 48–51
  market explanation for, 38–42, 41f
  statutory legislation on, 26–27, 30–38, 45, 49, 183, 188n1
  trade unions' influence on, 17–24, 27n11, 43–44, 44n18, 48–49, 182–184
Hobsbawm, E. J., 20

Index

Hoover, Herbert, 18n5
hours-earnings relationship, 151–181
　input variables and, 152–156
　"labor supply" as synonym for "hours," 164–169
　output relationship to, 151–152
　for self-employed workers, 172–180, 174–177nn10–16, 176t, 178f
　substitution effect vs. income effect, 153–156
　supply-and-demand hybrid of, 169–172
　supply vs. demand identification assumptions, 156–163, 164–165f
　trade unions' effect on, 180–181
"Hours of Work and Output" (Kossoris & Kohler), 104–105, 206
hours-output relationship
　Craig's observations on, 116–118, 117t, 118–119f
　elasticity of output, defined, 50, 60–62, 60n2
　in First World War munition factories, 72–84, 73–75t, 75–76f, 77t, 78–80f, 81t, 82–83f. See also Vernon case study
　IHRB on, 96–104, 97f, 98–99t, 100–101f, 102t, 103f
　Kossoris and Kohler's observations on, 106–109, 107–108t
　at macroeconomic level, 137
household well-being, 149–150
Huberman, Michael, 42, 181
Hunnicutt, Benjamin K., 22, 29

IHRB. *See* Industrial Health Research Board
Illinois, legislation on work hours in, 35
income effect, 153–156. *See also* hours-earnings relationship
Independent Labour Party (Britain), 45
Industrial Health Research Board (IHRB) case study, 95–104
　data collection methods, 96
　data used for, 203–205
　diminishing returns to hours conclusion for, 126–132, 130f
　on hours-output relationship, 96–104, 97f, 98–99t, 100–101f, 102t, 103f
　Second World War munition factories, overview, 95–96
Industrial Workers of the World (IWW), 46
Ineson, Antonia, 139n3
injuries, job-related, 148–149
intensity of work. *See* work effort
Iron and steel sector, reduction of work hours in, 22

Jones, Ethel B., 37

karoshi (death from overwork), 147n7
Kaufman, Roger, 56
Keane, Michael P., 165n8
Kellogg, W. K., 28, 29, 189
Killingsworth, Mark R., 165n8
Kivimaki, Mikal, 146
Kodz, Jenny, 7, 35

249

Kohler, Reinfried. *See also* Kossoris and Kohler case study
"Hours of Work and Output," 104–105, 206
Korea, working hours legislation in, 148
Kossoris, Max. *See also* Kossoris and Kohler case study
early research on absenteeism, 105n17
"Hours of Work and Output," 104–105, 206
Kossoris and Kohler (K&K) case study, 104–111
data collection methods, 104–105, 125–126
data used for collection methods, 206–210
on hours and absenteeism, 109–111, 109t
on hours and output, 106–109, 107–108t
on work injuries, 105n18
Kosters, Marvin, 156–159
Krusell, Per, 42n17
Kuhn, Peter, 7
Künn-Nelen, AnneMarie, 134

labor unions. *See* trade unions
laissez faire principles
historical reduction in work hours and, 43
on statutory legislation of work, 32
law of diminishing returns. *See also* hours-earnings relationship; hours-output relationship; production function
conceptual framework defined, 2–3
demand for hours vs., 40–41
implications of, 186–187. *See also* health issues of workers
for overtime work, 136–137
for part-time vs. full-time work, 133–136
unit returns to hours vs. constant returns to hours, 52–53
least-square estimates
Craig's observations on, 116–118, 117t
of days of work and output, 86, 87t
demographics and long weekly working hours in America, 7, 8t
hours-output relationship, IHRB on, 97–98, 99t, 100–101, 102t
hours-output relationship, K&K on, 108, 108t, 110
hours-output relationship, Vernon on, 70–71, 74, 77, 77t
including vs. omitting square of hours, 128–130, 129–131f
of recovery from work, 91t, 93, 94t
for understanding output-hours relationship, 66–67, 185
length of working hours
in contemporary economy, 5–12, 6t, 8–10t, 168
marginal product and, 152
part-time vs. full-time workers, 133–136
"short" vs. "long" hours, 79–80, 95, 100–102, 102t, 126, 155–156, 186
Lester, Richard A., 17, 20, 166–167

Lethbridge, Lucy, 14
Lever, William Hesketh, 28, 29
Lewis, H. Gregg
   Cobb-Douglas production function and, 59
   on history of reduction in work hours, 23
   Kosters' research and, 156, 156n3
   on market equalizing wage curve, 159–160
   on reduction in hours vs. rise in pay, 39–42, 40n16, 41f
Lloyd George, David, 69
"long" hours. *See* length of working hours
Loveday, Thomas, 86
Lowe, Rodney, 28
Lozano, Fernando, 7
lump of labor hypothesis
   Craig's observation on, 119–122, 120f, 121n21
   defined, 5

macroeconomics, hours-output relationship for, 137
MaCurdy, Thomas E., 158–159
manufacturing work hours, 20, 22, 31–37, 42, 153n2
marginal product of working hours
   CD-1 and CD-2 and, 61, 61f, 78, 79f
   Craig's observations on, 116, 117t, 118, 119f
   defined, 2
   GZ-2 on, 81–88, 82f
   length of working hours and, 152
   of overtime work, 136–137
   of part-time vs. full-time work, 133–136

market-wage function (market equalizing wage curve), 159–160
Marmot, Michael, 143
Marshall, Alfred, 54
Massachusetts, legislation on work hours in, 35, 36
Mather, William, 25–27, 189, 190
Matthews, Robin C. O., 21, 188
McCulloch, John Ramsay, 33n14
McGregor, Peter G., 136
medical workers, health issues of, 141–142
Mills, Harry A., 22
Minns, Chris, 42, 181
Missouri, legislation on work hours in, 35
Montgomery, Royal E., 22
munition factory case studies. *See* Industrial Health Research Board case study; Vernon case study

National Association of Manufacturers, 188n1
National Board of United Cloth and Hat Makers, 44
National Longitudinal Survey of Youth, 146–148
New Jersey, legislation on work hours in, 35
New York, legislation on work hours in, 35
Newman, George, 69
nominal hours vs. effective hours, 54–57, 185–186

Index

Oi, Walter, 138
opting out, work hour limits and, 34
output. *See* hours-output relationship
overtime pay
   augmented production functions for overtime work, 136–137
   early formalization of work hours and, 16
   historical reduction in work hours and, 49
   manufacturing industries, 153n2
Owen, Robert, 26–27, 189
ownership, capitalist vs. co-operative. *See* co-operative ownership; Craig case study

Paget, James, 140
Panel Study of Income Dynamics, 145–146, 179
Parker, Simon C., 179
Parry, David M., 188n1
part-time vs. full-time work, 133–136
Pencavel, John, 124, 154, 180n17
Peninsula Plywood, 124n22
piece-rate methods vs. time rates of pay, 54–57. *See also* Vernon case study
plywood mills. *See* Craig case study
police officers, health issues of, 141–142
Proctor, Susan P., 144
production function conceptual framework, 52–67. *See also* case studies of production functions
   Cobb-Douglas production function and, 57–62, 61f
   constant returns to hours and, 51
   defined, 1–2
   Gompertz and, 62–67
   plant's production function as conduct of employer and employees, 66–67
   unit returns to hours vs. constant returns to hours, 52–53
   work effort and, 54–57
productivity, part-time vs. full-time, 134
putting-out system, 15–16, 19

railroad employees, reduction of work hours for, 37–38
Rand Report (1966), 156
religion and reduction in work hours, 29–30, 45–46
returns to scale, defined, 50n20
Reynolds, Lloyd G., 166
Rhyatt, Rosina, 139
Ricci, Judith A., 140
Roosevelt, Franklin D., 18n5
Rosen, Sherwin, 41–42, 160
Rosenzweig, M. R., 56
Rowntree, Joseph, 189
Rowntree, Seebohm, 29–30, 189, 191
Russek, H. I., 144

Salford Ironworks, 25
Sauermann, Jan, 135
Schneiderman, Rose, 44
Schumpeter, Joseph A., 3
Scotland, Nigel, 45
self-employed workers, 172–180, 174–177nn10–16, 176t, 178f
seven-day working week, 84–88, 87t

Sharpless, Rebecca, 14
Shepard, Edward, 137
Shetty, Kanaka D., 142
Shirvani, Alex, 34–35
"short" hours. *See* length of working hours
short-time compensation, 18–19n6
Smith, Adam, 24, 24–25n8
Smuts, Robert W., 43
standard of living, working hours vs. , rise in pay, 1, 38–39
statutory legislation. *See* America; Britain
Stigler, George J., 3
substitution effect, 153–156. *See also* hours-earnings relationship
Sunday work
 additional pay for, 107
 holidays vs., 85n8
 seven-day working week, 84–88, 87t
supply of workers' hours. *See also* hours-earnings relationship
 identification assumptions about, 156–163, 164–165f
 "labor supply" as synonym for "hours," 164–169
Supreme Court (US) on work hours, 36

tax rates, 1
taxi drivers, hours-earnings relationship for, 178
Thatcher, Margaret, 47–48
Thom, Deborah, 139n3
Thompson, Nannie, 14

time rates of pay vs. piece-rate methods, 54–57
time-series movement research. *See* Lewis, H. Gregg
TNT poisoning, 139
trade unions
 history of working hours and influence of, 17–24, 27n11, 48–49, 182–184
 hours-earnings relationship and, 180–181
 membership increase from 1897 to 1913, 43–44, 44n18
 translog production function, 62n3
 Treble, John G., 55–56
 truck drivers, health issues of, 141–142
Typographical Association, 44

unemployment, 40n16
unions. *See* trade unions
unit returns to hours of work
 constant returns to hours vs., 52–53
 defined, 30, 50
 proportional changes in hours vs. output and, 95
 reduction in hours vs. rise in pay, 40
 using Cobb-Douglas convention function vs. augmented function for, 58–62, 131–132
United States
 Adamson Act (1916), 37
 agricultural and domestic work in, pre- and post-Industrial Revolution, 13–16
 average weekly hours of work in, 6–7, 6t, 8t, 9–12, 168, 181

253

United States (*cont.*)
  Bureau of Labor Statistics, 150
  Census of Population, 6–7, 6*t*, 8*t*, 156, 161
  Current Population Survey, 23, 163
  Department of Labor, 70, 104, 206
  Fair Labor Standards Act (1938), 38, 149
  historical reduction of work hours in, 24, 27–38, 42–48, 182–184
  Hours of Service Act (1907), 37
  majority of workers as employees in, 153
  National Industrial Conference Board, 70
  National Industrial Recovery Act (1933), 18n5
  National Recovery Administration (1933), 18n5
  Occupational Safety and Health Act (1970), 141
  Supreme Court on work hours, 36
  wage-earnings relationship in, 167
  Walsh-Healey Public Contracts Act (1936), 38
  working hours in, overview, 3–4

vacation time
  historical reduction in work hours and, 43
  "no vacation nation," United States as, 10
Van Buren, Martin, 37
Van Soest, Arthur, 168
Vernon, Horace, 70–72
Vernon case study, 68–95
  on absenteeism, 70–71
  conclusions from, 94–95
  data collection methods, 70–72
  data used for, 193–202
  Denison on typical hours of work and, 167
  diminishing returns to hours conclusion for, 126–132, 129*f*
  on fatigue and recovery from work, 88–94, 90–91*t*, 93–94*t*
  on hours-output relationship, 72–84, 73–75*t*, 75–76*f*, 77*t*, 78–80*f*, 81*t*, 82–83*f*
  on seven-day working week and, 84–88, 87*t*
  First World War munition factories, overview, 68–70
Virtanen, Marianna, 143–145

Washington (state), legislation on work hours in, 36
White, Michael, 150
Whitehall I health study, 142–143
Whitehall II health study, 142–144
Wilson, Lilian E., 139
Wisconsin, legislation on work hours in, 35
Wolman, Leo, 27n11
women. *See* gender issues
Woolwich Arsenal, 26
work effort
  effective hours of employees and, 4, 53–57
  historical reduction in work hours and, 43
  hours-earnings relationship and, 157, 167–168
  as input of labor, 54

Index

work injuries, 148–149
work sharing, 18
workers. *See* employees;
   self-employed workers
working hours, 1–12. *See also* length
   of working hours
   in Britain and America,
     overview, 3–4
   effective working time vs., 4–5
   law of diminishing returns
     and, 2–3
   production function of, 1–2
   standard of living and, 1

worker-hours, defined, 5
Working Time Directive
   (EU, 1998), 34–35
worklife balance, 149–150
Workplace Employee
   Relations Survey
   (Britain), 7–9, 9*t*
Wright, Gavin, 43

Yao, Xiaoxi, 146

Zeiss Optical Works, 26, 70
zero-hour contracts, 48n19

www.ingramcontent.com/pod-product-compliance
Ingram Content Group UK Ltd.
Pitfield, Milton Keynes, MK11 3LW, UK
UKHW021249180426
11946UKWH00003B/34